Place Names
of the Falkland Islands

Richard Munro

Published for the Shackleton Scholarship Fund
by Bluntisham Books

First published in 1998 by Bluntisham Books,
Oak House, East Street, Bluntisham,
Huntingdon, Cambs PE17 3LS
for Shackleton Scholarship Fund, Stanley, Falkland Islands

© Text and drawings R. Munro 1998

Map on front cover reproduced from Ordnance Survey map with the permission of The Controller of Her Majesty's Stationary Office, © Crown copyright MC88042M

ISBN 1 871999 09 X

Printed in Great Britain by Victoire Press, Bar Hill, Cambs

Introduction

Place names of the Falkland Islands date from the first sighting in 1592. Since then the early navigators, military garrisons, whalers, farmers and ships have all left their names for posterity. Indeed, the history and culture of the Islands are inextricably linked to the names of its topographical features, settlements, streets and houses.

This small dictionary of place names is by no means exhaustive. The names are derived from archives, maps, charts and conversations with Islanders and others whose interest and affection for the Islands match mine. I have omitted many names whose derivations are obscure and, naturally, I have attempted to be as accurate as possible. However, the study of place names is dynamic and suggestions for future editions would be welcome.

The assistance of the many individuals both in the United Kingdom and in the Falkland Islands who encouraged, informed and corrected me is especially acknowledged. In particular, the patrons and staff of the splendid Falkland Islands Museum deserve the greatest praise. I am also indebted to the late Robert Boumphrey, a family friend, for giving me permission to use those derivations that he recorded in 1950. Above all, I am extremely grateful to the committee of the Shackleton Scholarship Fund who had the foresight and courage to allow me to embark on the project.

I hope that reference to this little book will promote a deepening interest in these fascinating and historic Islands.

Richard Munro
Beyton, Suffolk

Spanish Topographical Terms Used:

Arroyo	Stream
Cerro	Hill
Isla	Island
Laguna	Pool
Monte	Mountain
Morro	Bluff
Rincon	Corner
Rio	River
Roca	Rock

Map References

The figures in brackets after names are full UTM grid co-ordinates (Zone 21). Most of the place names will be found on 1:250 000 and 1:50 000 scale maps of the Islands.

PLACE NAMES OF THE FALKLAND ISLANDS

ADAM, MOUNT [TC8781] Probably after Admiral Sir Charles Adam KCB MP (1780-1853), C-in-C West Indies 1841-45; a Lord of the Admiralty 1835-41, 1846-47.

ADELAIDE [TD9003] Possibly after the wife of William IV or the town in Australia, itself named after her.

ADVENTURE HARBOUR [UC5816] ISLAND [UC6116] SOUND [UC6325] After the survey sloop HMS *Adventure* (Cdr King) surveying the Falkland Islands 1826-30; or the schooner *Adventure* (formerly the *Unicorn)*, bought by Captain Fitzroy of the *Beagle* from William Lowe of Port Louis in 1833, and named after the first *Adventure*.

Acronyms and Abbreviations

DAP These letters are to be seen around Stanley painted in red or white within a crossed circle on various stone houses. They stand for *Defensa Area Pasiva* (Passive Defence Area, i.e. safe/non-combatant houses). The sign was painted on houses and buildings by civilians on the advice of the Argentine military authorities during the occupation of 1982.

FIBS Falkland Islands Broadcasting Station.

FIC Falkland Islands Company

FIGAS Falkland Islands Government Air Service

FIPASS [VC4473] Floating Intermediate Port And Storage System - a large floating jetty and warehouse structure commissioned in 1984.

MPA [VC0058] Mount Pleasant Airport, where the majority of the military garrison is located.

AIGUADA COVE [VC2390] Possibly from Spanish *aguada* (water supply).

AJAX BAY [UC5685] After the cruiser HMS *Ajax* which anchored in the bay on 28 November 1938.

ALBATROSS ISLAND [TD4303] After the black-browed albatross, also called Molly Hawk or Molly Mawk.

BLACK BROWED ALBATROSS

ALBEMARLE HARBOUR [TC5718] RINCON [TC5814] ROCK [TC6909] PORT [TC6513] Probably after George Keppel, 3rd Earl of Albemarle (1724-72), brother of Augustine, Viscount Keppel (see Keppel Island).

ALLARDYCE STREET [Stanley] After William Lamond Allardyce CMG, Governor of the colony 1904-15.

ALICE, MOUNT [TC5416] Possibly after the wife of T.F. Callaghan CMG, Governor 1876-80, or one of the Felton family.

AMES RINCON [UD0002] Ames was the name of a horse that died here in the early 1900s.

ANCHOR INLET [TC4814] ISLAND [UC2012] Descriptive.

ANNIE BROOKS BAY [TC9238] After the 96-ton inter-island schooner, commanded by Captain Smithers, wrecked here on 18 April 1874.

ANSON [VC2290] Named in 1842 after Admiral of the Fleet Lord Anson (1697-1762) who recommended the occupation of the Falkland Islands after his voyage round the world 1740-44. Now Port Louis Settlement.

ANXIOUS PASSAGE [UD0405] Presumably the passage was difficult to navigate.

ARARAT, MOUNT [TD4105] After the mountain in Turkey on which Noah's Ark is reputed to have settled as the Flood receded.

ARCH ISLANDS [TC6108] Descriptive of the natural rock arch to be seen at the islands.

ARCHIE'S POND [UC5943] Archie Short was a manager on Bleaker Island.

ARMANTINE BEACH [VD1606] After the French ship *Armantine* wrecked here on 21 August 1851. Also known as French Wreck.

ARROW HARBOUR [UC6652] ISLAND [UC6852] POINT [VC4576] After the surveying ketch HMS *Arrow*, stationed at the Falkland Islands from about 1839-43. Her Commanding Officers were Commander B.J. Sulivan, Lieutenant John Tyssen and Lieutenant W. Robinson.

ARROYO MALO [TC8338] (Spanish) Bad Stream. So called because of the difficulty in crossing the river on horseback or with livestock before bridges.

ARTHUR, MOUNT [UC0989] Possibly after Arthur Barkly, Governor 1886-87, and [TC7788] possibly after Arthur Felton.

ARTHUR'S PASS [TC7589] Arthur Felton once chased bulls here.

ARTHUR'S GATE [UC7292] After Arthur Alazia, who worked at Port San Carlos for many years. He was the radio operator at one time and also classed wool.

BABAS, LAGUNA [UC6865] (Spanish) Slimy Pool.

BABY, THE [UC0380] The feature is reputed to look like a baby's bottom.

BACK BEACH [UC3102, UC8132] Descriptive.

BAKER'S ROCK [UD3501] Possibly after a Dr Baker, Government Geologist. He was appointed in 1920.

BALD ISLAND [TC2757] Probably descriptive.

BALL MOUNTAIN [UC9081] Descriptive.

BALLAST BEACH [TD5112] Ships frequently took up ballast from pebble beaches.

BALLION STREAM [UC2175] Properly Baillon, after A.E. Baillon, Port Howard sheep farmer and manager of Falkland Islands Company 1890-98.

BARCLAY ISLAND [TC1756] After the *Barclay*, a Nantucket or New Bedford (USA) ship connected with the sealers and whalers who operated in the area.

BARRACK STREET [Stanley] So named because it leads to the barracks that were erected in 1849 to house the newly arrived Greenwich and Chelsea pensioners. They were subsequently occupied by a detachment of Sappers and Royal Marines in 1858.

BARREN ISLAND [UB1793] REEFS [UB2197] Descriptive.

BARROW HARBOUR [UC5420] After Sir John Barrow, Bart (1764-1848), Secretary of the Admiralty 1804-06 and 1807-44.

BARTON ISLAND [UD0008] After A.G. Barton, manager of Pebble Island in the 1930s.

BASIN, THE [TC8845, TC6241] Descriptive.

Battles

On 8 December 1914, the most decisive battle in the South Atlantic was waged between the British and German naval fleets. Street names in Stanley - Canopus, Glasgow and Kent - bear witness to the part that these three warships played in the hunting down and eventual destruction of their opponents. In 1938, HMS *Ajax* was anchored for a while in San Carlos Water and gave her name to Ajax Bay. She was later to take part in what was to be called the Battle of the River Plate and the eventual destruction of the German warship *Graf Spee*.

In 1982 an Argentine force invaded the Islands. The brilliant British combined operation that defeated them is commemorated in the names of brave commanders and soldiers - 'H' Jones Road, McKay Close, Jeremy Moore Avenue, Fieldhouse Close - and Thatcher Drive after the Prime Minister at the time. Of course, Mount Pleasant Airport, where the garrison is deployed, abounds with military-derived names.

BAY OF HARBOURS [UC4313] Allusive to the many sheltered coves in the bays.

BAY POINT [UD0610] Topographical.

BEACH POINT [VC3365] Topographical.

BEACON POINT [TC2557] Where a beacon once stood.

BEAGLE RIDGE [VC3680] Named in 1943 after HMS *Beagle*, the surveying ship commanded by Captain Robert Fitzroy RN which was engaged in surveying the islands in 1833-34.

BEATRICE COVE [VC4878] An old name, perhaps after a wreck, but no record or local tradition of a wreck of this name can be traced.

BEAUCHÊNE ISLAND [UB5137] After Jacques Gouin de Beauchêne who, in the *Phelipeaux* with the *Maurepas* (Captain Terville) and the *St Louis* discovered the island in 1701.

BEAUFORT, MOUNT [TC8683] After Rear Admiral Sir Francis Beaufort KCB (1774-1857), Hydrographer to the Navy 1829-55. He was instrumental in obtaining the services of Charles Darwin for the voyage of the *Beagle* and selected Sir Bartholomew James Sulivan to survey the Falkland Islands.

BEAVER BAY [TC0649] HARBOUR [TC0848] ISLAND [TC0547] The *Beaver* was a whaling ship from New Bedford. In 1773 she was commanded by Captain Hezekiah Coffin and took cargoes of sperm whale oil to London. She was also one of the ships involved in the Boston Tea Party. In 1792, she was allegedly the first whaler out of an American port to round Cape Horn and enter the Pacific.

BEDSTEAD PASS [VC0865] A local name probably so called because a bedstead or part of one was left in the vicinity by someone moving to or from Hillside.

BEEF ISLAND [TC0561] Descriptive.

BENSE ISLAND [TC5589] Obscure; however, it is named on Edgar's chart of 1797.

BERKELEY SOUND [VC3886] So named by Admiral the Hon John Byron (1723-86) in 1765 during his voyage round the world. Probably after James Berkeley KG, 3rd Earl of Berkeley, Vice Admiral of Great Britain 1717-36 and First Lord of the Admiralty 1717-37. It was named 'Baye Francoise' and 'Baye d'Acaron' by the French under Bougainville, and 'Puerto de la Soledad' (Port Solitude) by the Spanish.

BERTHA'S BEACH [VC0550] DITCH [VC0352] The Norwegian iron barque *Bertha* was wrecked here in 1892 with a cargo of cedar, a useful resource.

BETTS' PADDOCK [UC1797] After Willy Betts, farm manager on Pebble Island in 1940s and 50s. He used to keep horses in this paddock after shuttling them over in a scow from the island.

BIG CAPE, POND, ISLAND, etc. Descriptive.

BILLY'S CREEK [TD9901] After a horse of this name.

BILLY ROCK [VC5174] Diminutive of Port William.

BINNEY'S DOUBLE STREAM [UC1666] GATE [TC9945] After Horace 'Su' Binnie, born 1916, a shepherd at Fox Bay.

BIRD ISLAND [TC3212] Faunistic.

BLACK HILL, ISLAND, POINT, etc. Descriptive, topographical.

BLACKBURN RIVER [TC9893] The stream is particularly dark due to peat.

BLACKFISH CREEK [TC5214] A stranding of Blackfish (long-finned pilot whales) occurred here.

BLACKLEY'S [UC6890] After John Blackley, a boundary rider from Douglas Station whose house was here. He and his wife had two pretty daughters who were much admired by the men from Port San Carlos. After John's death, William Keith Cameron gave his pipe-smoking wife free tobacco for life from the Port San Carlos store.

BLANCO BAY [VC4276] (Spanish) White ... Possibly so named because of the whiteness of the sand in the bay or after someone's surname. Named on a Spanish map of 1789.

BLEAKER ISLAND [UC7012] Properly Breaker Island due to the large waves that crash against it. Appears also as Long Island on the chart compiled by the *Beagle* survey 1834.

BLOW HOLE [UD6717] A hole on top of the sea cliff through which water spouts in rough weather.

BLUE BEACH [UC6086] A beach at San Carlos Settlement named after the code word given to it by the British Forces during the landings in 1982.

BLUE MOUNTAIN [UC0147] Descriptive.

BLUFF COVE [VC1966] Topographical.

BOB'S ISLAND [UC9798] CREEK [UC4297] After Bob Skilling, a shepherd at Douglas Station in the late 1800s and at Port Howard in the early 1900s.

BOCA HOUSE [UC6359] (Spanish) Mouth. At the head of Brenton Loch.

BODIE CREEK [UC6353] PEAK [UC6776] After a Mr Bodie, master of the surveying ketch HMS *Arrow* (vide Arrow Harbour) in 1842 and mentioned in Governor Moody's Despatch No. 5 of 5 March 1842.

BOLD COVE [UC2979] POINT [Various] Descriptive.

BOMBILLA [UC8089] (Spanish) (Pronounced locally Bombija). A South American pipe-like implement used for drinking *maté*, a Paraguayan tea.

BONNER'S BAY [UC5983] Jack Bonner was owner and farm manager of San Carlos in the 1940s and 50s.

BOSOM HILLS [TC8849] Descriptive (!)

BOUGAINVILLE [UD9610] CREEK [VC2391] CAPE [UD9815] ISLA [UC9937] After Louis Antoine de Bougainville FRS (1729-1811), Secretary to the French Embassy in London 1755; ADC to Montcalm in Canada 1756; founded the French settlement at Port Louis, Berkeley Sound, in 1764; surrendered it to the Spanish in 1766 and circumnavigated the world in the frigate *Le Boudeuse* 1766-69; served against the English in the American War of Independence.

> **Bougainville**
>
> Unbeknown to the British, the great French navigator, mathematician and soldier, Louis Antoine de Bougainville had landed in 1764 at the cove he was to call Port Louis and had claimed the Islands for France. His crew's names live on in Chabot Creek, Duclos Point, Mount Simon and his ship, *L'Aigle*, gave her name to Aguila and Eagle Point. It was de Bougainville who named Choiseul Sound after the French Minister of Foreign Affairs. His name survives also in the exotic plant bougainvillaea which he later discovered during his travels.

BOULDER POINT [TC0844] Topographical.

BOUNDARY HILL, HOUSE, STREAM, etc. Descriptive of the delineation between farms.

BOX ISLAND [UD0207] After the native Boxwood *Hebe elliptica*.

BRANDON ROAD [Stanley] After the Very Reverend Lowther Brandon, Dean of Christ Church Cathedral, Stanley, and also Colonial Chaplain 1877-1907. It was Brandon who, with the inspiration of Bishop Stirling, supervised the building of the new cathedral.

BRAZO DEL (LA) MAR [VD0605, UC7533] (Spanish) Arm of the Sea, Sound.

BRENTON LOCH [UC6160] Probably after Vice Admiral Sir Jahleel Brenton, Bart (1770-1844).

BRETT HILL [TD7606] HARBOUR [TD7806] After Admiral Sir Piercy Brett (1709-81) who, as a Lieutenant and later as a Captain of HMS *Centurion*, circumnavigated the world with Lord Anson, 1740-44, and was one of the Lord's Commissioners of the Admiralty.

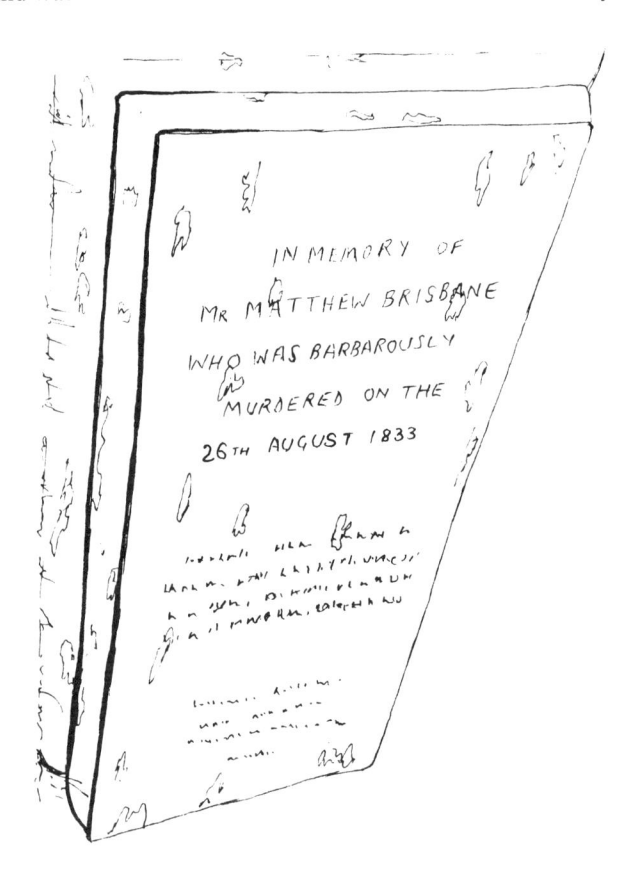

IN MEMORY OF MR MATTHEW BRISBANE WHO WAS BARBAROUSLY MURDERED ON THE 26TH AUGUST 1833

BRISBANE ROAD [Stanley] After Matthew Brisbane, an English sea-captain who served under James Weddell as master of the cutter *Beaufoy* and was in charge of the Port Louis settlement in 1833, when murdered by a band of convicts and gauchos. He is buried in the Port Louis cemetery.

Brown Flat, Harbour, Hill, etc. Descriptive.

Bull Flat Hill, Island, Point, etc. Faunistic.

Bull's Head [TC9550] A bull was killed at this point and its head left here.

Burnt Island [various] Probably so called because sealers often set alight the tussac islands to drive seals to the beaches.

Burntside [UC6661] Properly Burnside, i.e. beside a stream.

Bush Pass [VC1160] Descriptive.

Button Island [UC5928, TD6006] Descriptive of its small size or possibly after Jemmy Button, a Fuegian who spent time at the Keppel Mission. He had previously been one of the natives whom Fitzroy brought to England in 1830.

Byng, Mount [TD5113] Possibly after one of HMS *Carcass'* crew.

Byron Heights [TC5937] **Sound** [TC7595] After Admiral the Hon John Byron (1723-86) who commanded HMS *Dolphin* round the world 1765-66. He took possession of the Falkland Islands in 1765.

Cable Street [Stanley] The telegraphic cable that linked Stanley to Montevideo in Uruguay from 1915-21 terminated inside Cable Cottage, a house on this street.

Calf Creek [TD8501] **Island** [0201,1808] Faunistic.

Calista Island [UC0431] Possibly after the ship *Castallia*, a local schooner dragged ashore at Weddell Island in 1893.

Callaghan Road [Stanley] After T.F. Callaghan CMG, Governor 1876-80. When he retired a memorial was erected by public subscription as an appreciation of his services as Governor, in front of the then Roman Catholic Chapel in Dean St.

Calm Head [TC3118] Descriptive, an exposed southerly facing bluff where there is one spot alone where no wind can catch you. Few know where this spot actually is.

Camber, The [Stanley] A naval term for a small dock or man-made harbour. Situated on the north shore of Stanley harbour. Built during the 1890s as the Naval fuel depot to hold stocks of coal for ships of the South Atlantic Station. Tanks were built later to hold fuel oil. Between 1915 and 1922 a narrow-gauge railway linked it with the original Admiralty/Marconi wireless station at Moody Brook and so named because of its structure. It was given the local name of 'Klondyke' because of the high wages paid to the workers on its construction.

PLACE NAMES OF THE FALKLAND ISLANDS

CAMERON'S BROOK [UC6839] POINT [UC6291] RIDGE [UC6290] ROCKS [TC2851] After the Cameron family, owners of Port San Carlos.

CAMILLA CREEK [UC6466] Possibly from Spanish *camilla*, cot.

CAMPAMENTA BAY [VD2107] From Spanish *campamento*, encampment.

CAMPBELL DRIVE [Stanley] After Captain Ian Campbell, a popular Beaver float-plane pilot killed in 1976 when his aircraft crashed near Mare Harbour.

CAMPITO [UC5483] (Spanish) Small camp.

THE CANACHE

CANACHE, THE [VC4572] Corruption of 'careenage', a place where ships were 'careened', i.e. beached in order to have their bottoms cleaned at low water.

CANARD COVE [VC1988] (French) Duck ...

CANEJA CREEK [VC1291] From Spanish *caneca*, wooden vessel, bucket or possibly Spanish *conejo*, rabbit.

CANOPUS HILL [VC4572] After HMS *Canopus*, engaged in the Battle of the Falklands, 1914. She had an observation post on this hill at the time of the battle.

CANTERA HOUSE [UC6069] (Spanish) Quarry ... whence possibly came the stone for the *saladero*, salthouse, on the opposite side of the 'Boca' but more likely just the stone for Cantera House built in the mid-1860s.

CAPE MEREDITH [TC5204] After Sir William Meredith, Bart, a Lord of the Admiralty 1765-66 (died 1790).

CAPE ORFORD [TC2129] So called as early as 1766 probably after Admiral Edward Russell, Earl of Orford (1652-1727) Treasurer of the Navy 1689-99, First Commissioner of the Admiralty 1694-99, 1709-10, 1714-17; or after Robert Walpole, Earl of Orford (1676-1743).

CAPE PEMBROKE [VC5074] So called as early as 1766; probably after Thomas, Earl of Pembroke and Montgomery (1656-1732/33). First Lord of the Admiralty 1690-92, 1701-02, Lord High Admiral 1702, 1708-09.

CAPRICORN ROAD [Stanley] After the 380 ton barque of this name that suffered irreparable damage whilst trying to pass Cape Horn with a cargo of coal in 1882. She returned to Stanley and, after a period of use as a storage hulk, was finally scuttled to the west of the town in 1942.

CARANCHO BLUFF [TC2822] Carancho (Spanish: vulture) is the local name for the Crested Caracara, a large bird of prey.

CARAVAN PADDOCK [UD9203] Shepherds would live in 'caravans' - shanties on skids pulled by tractors - when engaged on mobile jobs like fencing.

CARCASS ISLAND [TD5113] After HMS *Carcass*, commanded by Captain Patison, which followed HMS *Jason* to Port Egmont in 1766.

CAREENING COVE [UC6393] An old name for the harbour at Port San Carlos, where ships were 'careened', i.e. beached in order to have their bottoms cleaned at low water.

CAREW HARBOUR [TC4834] Perhaps after Captain Carew of Stonington, an American whaler and sealer, whose ship was burned at Port Egmont in 1837: or after John Ewen Carew, mate of the snow (similar to a brig - Dutch) *'Queen Charlotte'* which called at the Islands in 1876.

CAROLINE, MOUNT [UC2490] After the wife of Colonel George A.K. D'Arcy, Governor 1870-76.

CARTHORSE POINT [UD0801] ISLAND [UD0802] Carthorses would have been over-wintered on the island and moved there by scow from the point.

CARYSFORT [VD3504] Properly Carysford. So named by the Hon John Byron after John Proby, First Baron Carysford (1720-72), a Lord of the Admiralty, 1757 and 1763-65.

CASSIES ISLAND [UC6208] See Cassard Point.

CASSARD POINT [UC6409] After the French sailing ship of this name that was wrecked here in June 1906.

CASTLE ROCK [various] Descriptive.

CAT ISLAND [UC5188] Probably faunistic.

Christ Church Stanley

CATHERINE, MOUNT [TC6385] After a daughter of Arthur Felton?

CATTLE GROUND, POINT, etc. Descriptive.

CAVADI RIDGE [UC7998] Pronounced locally 'caveda'. Possibly from a corruption of Spanish *curvado*, curved.

CELERY ISLAND [VC2488] Wild celery *(Apium australe)* is native to the Falkland Islands and covered extensive areas prior to the introduction of sheep.

CERITOS [UC7663] (Spanish) Small Hills; named by South American gauchos and correctly spelt *cerritos*.

CERRO MONTE [UC6991] Properly 'Cerro Montevideo'. It is said to have been named by a homesick Uruguayan as it looks like the hill of that name in Uruguay.

CHABOT CREEK [VC2694] Probably after either one of de Bougainville's men or a crew member of the *Uranie*.

CHAFFERS' GULLET [TC7016] After Edward Main Chaffers, Master of HMS *Beagle* 1831-36.

CHALLENGER, MOUNT [VC2470] Named in 1942 after the survey ship HMS *Challenger* which visited the Islands in 1876. A previous HMS *Challenger* which came to the Islands in 1834 brought out Lieutenant Henry Smith RN as officer in charge at Port Louis.

CHAMORO VALLEY [VD1503] (Spanish - *chamarro*) Bare, bald, i.e. lacking vegetation.

CHANCHO POINT [UC5292] (Spanish) Pig ... a reference to wild pigs that inhabited the islands.

CHAPEL ROCKS [UC6187] Descriptive of their shape.

CHARLES POINT [VC4676] Previously known as Rock Point, a topographical description, and as Galloway Point. It may perhaps be named after Charles Darwin, the famous naturalist, who visited the Islands in HMS *Beagle* in 1833, but more probably after Charles Melville, a naval rating who came in 1833 with Lieutenant Henry Smith RN on the latter's appointment by the Admiralty to administer the Islands. Charles Melville took his discharge from the Navy and became part-owner of the schooner *Montgomery*, carrying on a sealing business round the Islands. He rented Volunteer Rocks for sealing purposes, moved to Stanley and became Government Pilot. He died in 1876.

CHARTRES RIVER [TC9364] SETTLEMENT [TC8766] So called before 1850 after Dr William Chartres, surgeon in the survey vessel HMS *Philomel* 1842-45. The settlement was named after the river.

CHATA FLATS [UC9088] etc. (Spanish) Flat ...

CHATHAM HARBOUR [TC2748] After William Pitt, first Earl of Chatham (1708-78) on whose recommendation the settlement at Port Egmont was occupied in 1766.

CHEEKS' CREEK [TC6937] Local family name.

CHEROOGS POND [UC5145] Gaucho patois for cheroots.

CHICA, ISLA [UC6521] (Spanish) Little Island.

CHICO ARROYO [TC8143] (Spanish) Little Stream.

CHIMANGO VALLEY [UD7810] (Spanish) After the Chimango Caracara (*Milvago chimango*), a large bird of prey.

CHOISEUL SOUND [UC9045] (Pronounced locally 'chisel') After E. de Choiseul, Duc de Stainville, French Minister of Foreign Affairs at the time of de Bougainville's founding of his settlement at Port Louis, 1764. The name Baye Choiseul seems originally to have been applied by the French to Port William.

CHRISTINA BAY [VC4973] After a German ship *Christina*, wrecked on the nearby Maggie Elliott rock in 1880.

CHRISTMAS HARBOUR [TC8368] The area was surveyed by Sulivan on Christmas Day 1842.

CHRIS'S CAMP, PASS [UD7214] After Chris Andreasen, a Danish seaman, put ashore in 1907 from his ship the *Agnes G. Donahue*, a Nova Scotian sealing schooner, with beri-beri. In 1910 he was engaged to work at Port San Carlos and remained there for 37 years, first as foreman and later, for 14 years, as manager. He died in Stanley in 1962.

CHURCH VALLEY [VC3193] A parson, Charles Bull, was once lost here in the fog whilst out walking. He was Colonial Chaplain 1860-76.

CLAM BED POINT [VC4074] Faunistic. Previously named Goose Point, also faunistic.

FITZROY — COACH HOUSE

COACH HOUSE [VC1064] A Leyland Tiger single-decker bus, now used for storage

COBB'S PASS [UC5354] After Frederick E. Cobb, manager of Falkland Islands Company 1867-91.

COCHON ISLAND [VC4582] (French) Pig ... So named because of its humpbacked appearance.

COFFIN ISLAND [TC0659] Possibly after one of the many whaling captains active between the mid 18th and 19th centuries with the surname Coffin, several of whom operated in waters close to the Falkland Islands. See also Speedwell Island.

COLLIERS, THE [TC0954] At one angle, the rocks resemble a ship of this type.

COLORADO BAY [VC0598] BROOK [UC9673] PASS [UC5349] (Spanish) Red...

COMODA DITCH [UC9758] (Spanish - *comodo*) Useful, handy. A convenient watering place off the track.

CONCORDIA BAY [UD8816] Where the German barquentine of this name was wrecked on 17 August 1891.

CONGO HOUSE [UC2837] (Spanish) Negro ...

COOKE HILL [TC6682] After Lieutenant John Cooke RN who served in the survey sloop HMS *Adventure* until 1827.

CORNER PASS [VC2977] Descriptive, the ford being at a bend in the Murrell River.

CORTLEY HILL [VC4074] Correctly spelt 'Cautley' after Captain Henry Cautley RE who arrived at Stanley on 20 August 1881 with instructions from the War Office to report upon the defensive capabilities of the Islands as a coaling and refitting station for the Royal Navy and Mercantile Marine.

COUTTS HILL [UC6799] After James Magnus Coutts, a dentist who served in the Islands from 1925-31. One account is given that he once rolled down this hill in favour of walking, being so tired and unfit. Alternatively, his horse fell here once, throwing Coutts who suffered a broken leg.

COW BAY, ISLAND, PADDOCK, etc. Faunistic.

CRAIGIELA POINT [UB3699] After a ship of this name wrecked in the vicinity in December 1879.

CRATES, THE [TD8900] Wooden structures resembling crates have often been filled with rocks and sunk in bays to form anchors or strong points for fencing.

CROOKED INLET [TC7583] Descriptive of the arm of the sea in which the settlement lies.

CROUCHING LIONS, THE [TC5980] Descriptive.

CROZIER PLACE [Stanley] After Captain Francis Moira Crozier RN (1796-1848) who commanded HMS *Terror* in the British Antarctic Expedition, 1839-43.

CURLEW BAY [UC6591] Faunistic.

CUSHY'S HILL [UC7197] Possibly after a gaucho whose shanty used to be here, 'Cushy' being a distortion of his name: Huesey (José?).

CUTTER COVE [UC4464] A cutter is a type of boat.

CYGNET HARBOUR [UC2849] Probably faunistic - swans are not uncommon in this part of Lafonia. Called Swan Island Harbour in 1843.

D'ARCY, MOUNT [UC2191] After Colonel George A.K. D'Arcy, Governor 1870-76.

DAIRY PADDOCK ROAD [Stanley] There was once a Falkland Islands Company dairy in this paddock.

DAN'S SHANTY CREEK [VC1690] BROOK [VC1893] Dan Lane, or Lehan, was a boundary rider at the turn of the 19th/20th centuries.

DARWIN [UC6558] After Charles Darwin, the celebrated naturalist, author of '*The Origin of Species*', who visited the Falkland Islands in 1833 and 1834 in HMS *Beagle* and spent a night near the present settlement.

DAVY'S PADDOCK [UC6794] After Davy Stewart who was a shepherd at Port San Carlos in the early 1900s.

DAVIS STREET [Stanley] After John Davis, of the ship *Desire*, who first sighted the islands on 14 August 1592.

Dean, Mount [TC4122] **Street** [Stanley] After John Markham Dean who came to Port Louis as clerk to John Bull Whitington, 17 November 1840. Dean was afterwards a well known merchant in Stanley and owned Port Stephens. He died in 1888.

Death Valley [TC7868] D. Brockway, a passenger in the ship *Echo*, was killed here on Christmas Day 1849 by his own shotgun whilst trying to flush out a warrah (fox).

Death's Head [TC4898] A sailor was reputed to have fallen to his death here.

Devil's Steps [TC7269] A natural rock formation that looks like stairs. Folklore states that they are believed to have been cut by the Devil.

Diamond Cove [VC3689] **Rincon** [TC5633] Descriptive of the shape of the cove.

Diddle Dee Island [UD0402] Diddle Dee (*Empetrum rubrum*) is a woody, resinous shrub producing edible berries which grows extensively over large areas of the Islands.

Dip Paddock, Point, Ponds, etc. Where were once sheep dips.

Dirty Ditch [various] **Point** [TC1939] Descriptive.

Discovery Close [Stanley] After Captain Robert Falcon Scott's ship *Discovery* which was sold to the Falkland Islands Government and carried out research in whaling and marine biology during the 1920s and 30s.

Dish Cover Hill [TD8811] Descriptive.

Docherty's Shanty [UC7977] Jimmy Docherty was a worker on James Greenshields' estate. He was drowned with his employer in a cutter that sank in Port Salvador in 1893 which had been loading the Falkland Islands Company schooner *Thetis* and was returning to Douglas Station.

Dockyard Islands [UD1507] Possibly where there was once a boat repair facility.

Doctor's Point [VC4376] **Creek** [TC9040] After Henry Joseph Hamblin, first Colonial Surgeon (see Hamblin Cove). In the early days of the colony there was a guiding beacon at Doctor Point with the arm pointing towards the Narrows of Port Stanley.

Doctor's Head [UC5791] Possibly named because it is near Hospital Point.

Doctor's Leap [TC9656] After a Dr Cunningham who crashed his Landrover here.

Doctor's Mount [UC0990] Where the doctor was met when travelling between Port Howard and Hill Cove.

DOLLY'S GATE [UD9207] After Alf May's wife, Dolly (née Biggs), from Douglas Station. On one occasion in the 1930s, whilst en route to Salvador, she opened this gate and her horse ran away.

DOLPHIN, CAPE [UD6222] Named after HMS *Dolphin* by Commander (later Admiral) the Hon John Byron in 1765. In this ship, in the company of HMS *Tamar*, he circumnavigated the world, 1765-66. He took possession of the Falkland Islands at Port Egmont on 12 January 1765.

DON CARLOS BAY [VC4879] After a local cutter '*Don Carlos*' wrecked here in 1889.

DONALD, MOUNT [TC8783] Possibly after Charles Donald, Surgeon in the *Active*, one of a four-ship exploratory whaling party from Dundee in 1892-93.

DOS LOMAS [UC4662] (Spanish) Two Hills.

DOTTEREL POINT [UC5992] Descriptive (a small bird).

DOUBLE CREEK ISLANDS, RINCON etc. Descriptive.

DOUGLAS STATION [UC8897] So named by James Greenshields, owner of the farm in the late 19th century after his wife - a Miss Douglas; or possibly after Douglas in Lanarkshire, the Scottish home of the Greenshield family.

DOYLE RIDGE [TC9154] MOUNT [TC8361] After Commander Charles Doyle, 1st Lieutenant in HMS *Philomel*.

DRIFTWOOD ISLAND [UC6307] Descriptive, for instance a ship's cargo of mahogany was once washed up in this area.

DRUNKEN ROCK [VC3477] A fanciful name, supposed to be 'one bottle's ride from Stanley' i.e. the distance one would go from Stanley before desiring liquid refreshment. The large number of empty bottles in the vicinity lends support to this definition.

DRURY STREET [Stanley] Perhaps after Captain Drury of HMS *Pandora* which called here in 1856; more probably after Lieutenant Drury of the detachment of Marines which arrived in Stanley in 1858.

DRY ISLAND [TD9804] POND Descriptive.

DUCLOS POINT [VC2990] After Captain Nicolas-Pierre Guyot, Sieur Duclos, who commanded the *Aigle*, one of de Bougainville's ships in 1765.

DUFFINS' BRIDGE [UC3523] After a Mr Harry Duffin, better known as 'Tom', shepherd at North Arm in the early 1900s.

DUNBAR [TC6295] Obscure but so called on a French map of 1771.

Rose Hotel, Stanley

DunNose Head [TC6461] Appears as 'Dunoze Point' on Edgar's 1797 chart, possibly after the headland of the same name on the Isle of Wight.

Duperry Harbour [VC2585] After L.I. Duperry, an officer in the French corvette *Uranie*, wrecked at the Islands in 1820. Later, in the *Coquille*, he took observations at the Falkland Islands, in 1822, probably on Hog Island.

Dutchman's Brook [VD3606] **Island** [VD3806] **Point** [TD9202] Possibly allusive to Ieergen Dettleff who was a German. See Hadassah Bay.

Dyke Island [TC3332] **House** [TC3534] **Point** [TC2039] Descriptive of the many ditch-like geological features in the area.

Eagle Point [VC4789] From the French *Pointe de l'Aigle*. So called by the French under de Bougainville after the frigate *Aigle* (eagle), one of the ships in which they travelled to the Falkland Islands in 1764. It appears as Nelson's Point on Wyld's plan of 1839 and was called Cape de Barra by Benjamin Morrell.

EAST BAY, COVE, ISLAND etc. Descriptive.

EDDYSTONE ROCK [UD5627] So named by Commodore (later Admiral) the Hon John Byron in 1765 probably after the rock of the same name off Cornwall.

EDGAR, PORT [TC7733] After Lieutenant Thomas Edgar RN who surveyed at the Falkland Islands in 1786-87.

EDGEWORTH, MOUNT [UC0187] Probably after the author Richard Lovell Edgeworth (1744-1817) of Edgeworthstown, Ireland. He was brother-in-law of Sir Francis Beaufort. See Mount Beaufort.

EGG HARBOUR [UC3652] Probably so named by reason of a plentiful supply of eggs having been obtained here by one of the surveying vessels, there being large gentoo penguin rookeries in the vicinity.

EGMONT, PORT [TD8706] So named by the Hon John Byron when he took possession of the Falkland Islands in 1765, after John Perceval, second Earl of Egmont (1711-70), First Lord of the Admiralty, 1763-66.

ELEPHANT BAY, BEACH, ISLAND, etc. Probably relating to elephant seals.

ELEPHANT CANYON [VC2569] To refresh themselves whilst working at this borrow pit for the new Stanley–Darwin road in the 1980s, workers drank copious quantities of Carlsberg 'Elephant' beer.

ELIZA COVE [VC4169] CRESCENT, ROAD [Stanley] Probably connected with the barque *Eliza* which visited the Islands in 1849 and was afterwards engaged by Samuel Lafone to run between Montevideo and the Falkland Islands.

ELLEN MOUNT [TC3824] Possibly after a member of the Dean family.

ENDERBY POINT [UC9335] After the 19th century London shipping and mercantile firm Messrs Enderby who were largely engaged in the whale and seal fisheries in the Southern Ocean.

ENDURANCE AVENUE [Stanley] After HMS *Endurance*, the ice patrol ship.

ESTANCIA HOUSE [VC1877] MOUNT [VC2079] (Spanish) Farm ...

EVANS SHIRT [UC2278] A surveyor named Evans once lost his shirt here.

EVELYN HILL [UC9782] So named by John Felton, owner of Teal Inlet Farm (originally named Evelyn Station), after his daughter Evelyn.

EVERGREEN FLAT [UC1984] Descriptive.

EXMOOR GULLY [UC1880] After the breed of sheep.

FAIRY COVE [VC3973] After a local schooner, the *Fairy* (104 tons), which may have run aground here at some time. She was laid down in 1850, bought by the Falkland Islands Company in 1853 and ran the mails between Stanley and Montevideo for twelve years. After eight years in the ownership of W. Bertrand and J. Switzer, West Falkland settlers, she had a varied career with the Falkland Islands Company until being broken up in 1932.

FALKLAND ISLANDS From Falkland Sound. The name Falkland's Land was first used by Captain Woodes Rogers who visited the Islands in 1708. Falklands Islands was the name given by the Hon John Byron when he took possession of them in 1765.

FALKLAND SOUND [UC3874] So named by Captain John Strong of the *Welfare* who navigated the Sound in 1690, after Anthony Carey, fifth Viscount Falkland (1656-94) who was Treasurer of the Navy 1681-89 and First Lord of the Admiralty 1693-94. It was named Carlisle Sound by the Hon John Byron in 1765, perhaps after Charles Howard, third Earl of Carlisle (1669-1738) who was First Lord of the Treasury 1701-02 and 1715.

FANNING'S HARBOUR [UC5594] HEAD [UC5195] ISLAND [UC5694] After Captain Edmund Fanning, a well known American sealer, who frequented the Falkland Islands in the late 18th and early 19th centuries. Author of '*Voyages Round the World*', New York 1833.

FAR PEAKS [TC7986] Descriptive.

FASCINE VALLEY [TD7116, TC9672] Properly *fachine* or *faschine* (*Chiliotrichum diffusum*) a tall, bushy plant that, whilst still fairly common now, used to be abundant before heavy grazing.

FEGEN INLET [TC3424], MOUNT [TC7289] After Lieutenant Charles Goodwin Fegen, Mate and Assistant Surveyor in HMS *Philomel* 1842-44.

FELTON STREAM [VC3772] After Sergeant Major Henry Felton, late of the Life Guards, who was Warrant Officer in charge of the military Pensioners at Stanley; born in 1797, he arrived in the Islands in 1849. Patriach of a well known Falkland family, his son Arthur farmed West Point Island for 50 years and another son, J.J., founded Teal Inlet.

FIELDHOUSE CLOSE [Stanley] Admiral Sir John Fieldhouse was the UK supreme commander of the operation to re-take the Islands in 1982.

FINDLAY CREEK [UC6151] HARBOUR [UC2235] After Alexander George Findlay (1812-75) the well known geographer and hydrographer.

FINDLAYS ROCKS [UC5994] After Findlay McLennan, a Port San Carlos shepherd of the early 1900s. He always said that he would like a house built here.

FISH CREEK, HOLE, POND, etc Faunistic.

FITZROY RIVER [VC0565] PORT [VC2163] ROAD [Stanley] SETTLEMENT [VC1561] After Captain (later Vice Admiral) Robert Fitzroy (1805-65) in command of HMS *Beagle* surveying on the coast of South America and at the Falkland Islands, 1828-36. He later became Governor of New Zealand 1843-45. He was subsequently noted for instigating the system of storm warnings which developed into daily weather forecasts.

FIVE SHILLING PASS [UC5254] Stockmen frequently held 'sweep' races when riding in Camp. In this case the takings would be five shillings to the winner.

FLAT ISLAND, PADDOCK, PASS, etc Descriptive.

FLORES HARBOUR [UC2311] Possibly after Luciana Flores, one of the murderers of Matthew Brisbane at Port Louis 26 August 1833; or after Colonel Flores who was killed at the Siege of Montevideo during the Parana Campaign, 1845-46.

FOAM CREEK [VC0796] After the schooner *Foam*, 65 tons, originally owned by Lord Dufferin and registered at Waterford, 1852. She arrived in the Islands on 23 October 1863 under Captain Smithers as the Government Mail and Pilot Vessel and was wrecked at Carcass Island in May 1890.

> **British Explorers**
>
> John Davis of the *Desire*, the first to record a sighting of the Islands in 1592, is commemorated in Davis Street, Stanley. Captain John Strong of the *Welfare* named the passage between the two main islands Falkland Sound after the Treasurer of the Navy whilst Captain John Byron, taking possession of the Islands in 1765, used the name for the whole archipelago. As well as Byron Heights he was instrumental in naming Fox Bay (after the local fox - warrah), Cape Dolphin (after his ship), Port Howard (Earl of Carlisle, the First Lord of the Treasury) and Port Egmont, after the Earl of Egmont, First Lord of the Admiralty. In 1833 Captain Robert Fitzroy, in command of HMS *Beagle*, called at the Islands. This visit is reflected in such names as Mount Kent, Port William, Hammond Cove and Darwin, after the eminent naturalist Charles Darwin who accompanied Fitzroy. A Lieutenant in the ship, Bartholomew Sulivan, later returned in command of HM Survey Ships *Arrow* (1838) and *Philomel* (1862) and settled in the Islands for a while building the original Sulivan House in Stanley. MacKinnon's Creek, Mounts Doyle and Richard, and Philimore Island were named at these times whilst his ships are remembered in many locations throughout the Islands.

FORESIGHT HILL [UD3204] Properly Forsythe, an injured survivor of the wreck of the *Lotus*, a local schooner owned by the Falkland Islands Company wrecked in Tamar Pass in 1872. He was left here whilst help was sought. Gauchos returned to the spot but, ominously, reported that they were unable to find him - or the money-belt he was known to be wearing!

FORTUNA POINT [UC2562] After the 163 ton schooner bought by the Falkland Islands Company in England in 1894. She was wrecked on West Island in Falkland Sound on 19 May 1906.

FOUL BAY [UD6311] A poor anchorage where currents can drag vessels aground.

FOX BAY [TC9136] Faunistic, after the Falkland Island fox *Dusicyon antarcticus* (called locally the 'warrah'), possibly the only quadruped indigenous to the Islands and now extinct. The bay was so named in 1765 by Hon John Byron (1723-86) who commanded HMS *Dolphin*.

FOX ISLAND [TC1857] After George Fox (1624-91) founder of the Society of Friends. See Friend Passage, Penn and Quaker Islands.

FREEZER [UC2378] Stock was held here prior to being transported to the freezer plant at Ajax Bay.

FREHEL, CAPE [VD1406] So named by de Bougainville after Cape Frehel, situated 'four leagues from St Malo'.

FRENCH PEAKS [TC8188] After M.C. French, Assistant Surgeon in HMS *Philomel*.

FRENCH WRECK [VD1606] See Armantine Beach.

FRIEND PASSAGE [TC1654] Allusive to the Society of Friends founded by George Fox (1624-91). See Fox, Penn and Quaker Islands.

FRIZZLEY BAY [UC1596] Local patois for 'cold'.

FRYING PAN, THE [VC0859] When travelling in the direction of Stanley, Peak Stream is shaped here like a frying pan.

FURZE BUSH PASS [VC3176] Descriptive, a furze bush (*Ulex europeus*) marking the ford over the Murrell River.

GALLINA ROCK [UC2072] (Spanish) Hen ...

GARDEN PASS, THE [TC9293] The people of Sound House had a garden here.

GARDINER'S PATCH [UD2105] The 88 ton schooner *Allen Gardiner* of the Keppel Island Mission once bumped on the kelp reef here in the 1850s.

GARNIER POINT [VC2385] Possibly after a member of the *Uranie*'s crew or de Bougainville's party.

GATEADO POND [UC4747] (Spanish) Probably refers to a horse with a coloured streak, known as a Gateo.

GEORGE ISLAND [UB1397] First appears after the survey made by HMS *Beagle*, perhaps after King George IV.

GEORGE LAMOSA HORSE POND [TC9653] After a much respected shepherd of pioneer stock at Port Howard (1893-1979).

GENESTA POINT [UD1918] After a local schooner owned by the Falkland Islands Company and wrecked near Port Egmont on 24 May 1888. She later drifted out to sea and was lost.

> **Gauchos and Cattle**
>
> From the days of first settlement until the mid nineteenth century cattle abounded on the Islands. Names such as Mocha, Cattle Ground, Rodeo Point bear witness. However, there is a greater incidence of names associated with those employed to round up and kill the beasts - Gauchos. Ceritos, Cushy's (possibly José's) Hill, Gregorio and Pony's Pass are examples. Strictly, the term refers to South American cattle hands. However, there were also many other nationalities employed as gauchos in the Falkland Islands including Scottish, English, French and Gibraltarian. A name made famous in the 1982 war, Tumbledown, derives from an early event when a herd of horses was driven over the steep rock face of the hill by gauchos.

GIBRALTAR ROCK [TD3807] STREAM [TC7224] So named by the well-known local family of Pitaluga, originally from Genoa, Italy. Of two brothers, one came to the Falkland Islands in the early 1840s and the other went to Gibraltar, where the family still exists.

GID'S ISLAND [TC7173] Gid McKay was a member of a large family that lived in the Chartres area in the 1920s and 30s.

GLADSTONE [UC3094] Probably after William Gladstone, Secretary of State for the Colonies in 1846, who later became Prime Minister.

GLASGOW ROAD [Stanley] After HMS *Glasgow* which took part in the Battle of the Falklands, 1914.

GOAT RIDGE [VC3171] So called after a flock of wild goats which used to graze in this locality.

GOLDING ISLAND [UD0907] So called as early as 1797, possibly after the prominent mound in the centre known as the Golden Ball.

GOLDSWORTHY ROCK [VC4173] After Sir Roger Tuckfield Goldsworthy KCMG, Governor of the Falkland Islands 1891-97.

GOOSE GREEN [UC6456] After the Upland Goose of the Islands.

> **Darwin**
>
> Whilst Charles Darwin (1809-82) was at Cambridge he was recommended as naturalist to HMS *Beagle* under Captain Robert Fitzroy who was to undertake a scientific survey of 'the South extremity'. During the six year voyage, Darwin collected much material for later publication in learned biological works. He is perhaps best known for his Theory of Evolution Through Natural Selection. Whilst at the Falkland Islands, rather than eulogise the stark beauty of the place, Darwin declared them to be 'miserable islands.....with a desolate and wretched aspect'. His name survives in Darwin settlement and, possibly, Charles Point.

GORING HOUSE [TC9663] After Goring-on-Thames, the English home of Frederick May, Senior, the first shepherd to live here.

GOSLING BAY, CREEK, etc Faunistic.

GRAND JASON [TD1335] See Jason Islands.

GRANNIE'S PASS [UC3596] Called thus after a carthorse of this name.

GRANTHAM SOUND [UC5175] After Thomas Robinson, Second Baron Grantham (1738-74), who recommended the abandonment of Port Egmont in 1774.

GRAVE COVE [TD4603] So named because of the sealers' and whalers' graves to be found there.

GRAVE POINT [VC2990] There are two graves here most likely to be those of E H Hellyer, Clerk of the *Beagle* (d. 1833) and Lieutenant Clive, HMS *Challenger* (d. 1834). They both drowned.

GREGORIO ROCK [UC6792] Probably after one of the gauchos who were recruited to eliminate the wild cattle.

GREY CHANNEL [TC1058] After Earl Grey [1799-1882) Secretary of State for the Colonies 1846-52; or after Captain the Hon George Grey RN of HMS *Cleopatra* at the Falkland Islands in 1836.

GULL ISLAND, POINT, etc Faunistic.

GUN HILL [TC8565] Descriptive - the rocks here can take on the appearance of a field gun from a distance.

GUTTERY PASS [UC7558] The land here is marked by several channels (gutters) in wet weather.

GWENDOLINE BAY [UC1326] Probably after the Falkland Islands Company's 108 ton schooner, hulked in the 1920s. She sank in Stanley harbour in 1966.

GYPSY COVE [VC4474] A fanciful name; a favourite spot for picnics.

HADASSAH BAY [VC4374] After the schooner *Hadassah*, owned by the Government, which carried mails to the West Falklands. She ran ashore in this bay whilst endeavouring to pass through the Narrows into Port Stanley on 1 November 1892. She later foundered at New Year Cove, Weddell Island, in October 1896. This bay was originally called Dettleff's Bay after Ieergen Christian Dettleff of Hamburg who went to the Colony in 1841 and became naturalised. See Dutchman's Brook.

HALFWAY ROCK [VC6699] Estimated to be half way between Port San Carlos settlement and Cape Dolphin house.

HAMBLIN COVE [VC4476] After Henry Joseph Hamblin, first Colonial Surgeon, who arrived at Port Louis on 25 November 1843 in the schooner *Colombian Packet*. He was for many years a member of the Executive Council and died at sea on a voyage to England on 23 June 1854 aged 55. He built Stanley Cottage. There is a memorial to him in Christ Church Cathedral, Stanley. See Doctor Point.

HAMILTON'S VALLEY [UD6519] After Dr J.E. Hamilton, Government Naturalist in the 1930s-50s, who spent a great deal of time on Cape Dolphin documenting the wildlife.

HAMMOND COVE [UC8840] After Robert N. Hammond RN "an early and much esteemed friend" of Captain Fitzroy. He was lent to HMS *Beagle* in November 1832 from HMS *David* of which he was then a mate, but returned to England in May 1833. He was also a friend of Admiral Sir Bartholomew James Sulivan.

Harps Farm [UC1185] Previously Plain House, the Smith family named it in 1986 by forming an acronym of their names: Heather And Robin, Patricia, Shula.

Harriet, Mount [VC3070], **Port** [VC3768] After the American sealing schooner *Harriet* (Captain Davison) which was one of those arrested by Vernet in 1831, the arrest leading to the bombardment and destruction of Port Louis by the US warship *Lexington*. Port Harriet was formerly called Port Firm, presumably descriptive of its quality as an anchorage.

Harriet, Mount [VC5617] After the wife of Mr Dean, owner at Port Stephens.

Harston, Mount [TD7213] After Captain Henry Harston RN who entered the Navy in 1826 and was Lieutenant in HMS *Philomel* 1842-45 during her Falkland Islands deployment.

Hawk Hill, Rock, (Nest) Valley, etc Faunistic, probably referring to the widely distributed Red-backed Buzzard (*Buteo polyosoma*).

Hearnden Hill [VC2190] **Water** [Stanley] After either Sergeant Robert Hearnden Royal Sappers and Miners or his brother Thomas, who both came to the Islands in 1841-42 with Governor Moody and party. Sergeant R Hearnden was appointed Superintendent of Public Works in 1844.

Hebe Street [Stanley] After the *Hebe* which brought Governor Moody and his party to the Islands in 1841-42.

Hell's Kitchen [TC2847] Topographical. Derived from a hole in the cliff side in which the waves "boil".

Henry Creek [TC6384], **Mount** [UC1291] Possibly after Lieutenant Henry Smith RN as officer in charge at Port Louis 1834-38 or (Mount Henry) Henry Waldron, the local land owner.

Herbert Stream [TC7496] After Herbert Felton, one of the first settlers on West Falkland.

High Bluff, Hill, Island, etc Descriptive.

Hill Cove, Gap, Head, etc Topographical.

H Jones Road [Stanley] After Lieutenant Colonel Herbert ('H') Jones VC, commanding officer of 2nd Battalion The Parachute Regiment in the 1982 war. He was killed at Goose Green and is buried at San Carlos.

Hog Island [VC2589] Where pigs were probably turned loose on this and other islands similarly named.

Hoggett Camp [VC1298] A hoggett (sometimes hogg) is a sheep aged under 12 months, i.e. prior to first shearing.

FIGAS

HOLDFAST ROAD [Stanley] Named after the military code word for Royal Engineers. They were ubiquitous in the Falkland Islands after the 1982 war.

HOLMESTED'S PADDOCK [TC9399] After Ernest Holmested, a pioneer who farmed the Shallow Bay area in the second half of the 19th century.

HOOKERS POINT [VC4672] After Sir Joseph Dalton Hooker (1817-1911) who was assistant surgeon and naturalist on the British Antarctic Expedition under Captain Sir James Clark Ross RN and visited the Islands in 1842.

HOPE COTTAGE, HARBOUR, etc After Captain Charles Hope RN (born 1798) commander of HMS *Tyne*, despatched with HMS *Clio* (Commander J.J. Onslow) in 1832-33 to re-take possession of the Colony which he did in December 1832.

HORNBY MOUNTAIN [UC1570] So named by Governor Moody in 1842. He gives no reason but perhaps referred to Rear Admiral Phipps Hornby CB, a Lord of the Admiralty in 1852.

HORQUETA VALLEY [UC6595] (Spanish) Fork ... The stream forks here. However, the valley was actually named after one of William Keith Cameron's (Port San Carlos) favourite horses.

HORSE ISLAND, PADDOCK, etc Faunistic.

HORSESHOE BAY [VC1093], THE [TC9885] Descriptive of the shape of the features.

HOSPITAL FLAT [TC0799] POINT [UC5892] Good grazing land near to farms where stud rams and sick animals were kept.

33

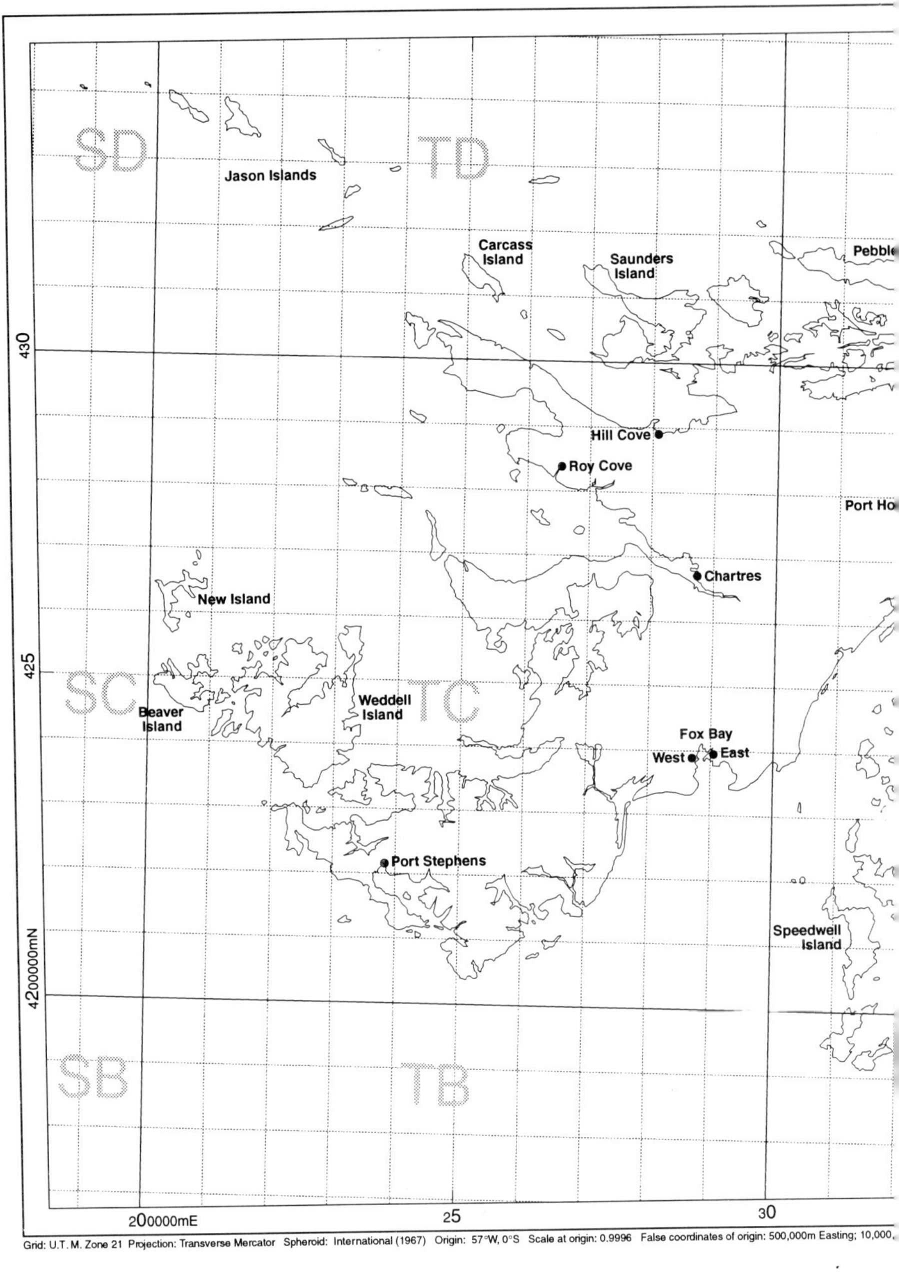

Grid: U.T.M. Zone 21 Projection: Transverse Mercator Spheroid: International (1967) Origin: 57°W, 0°S Scale at origin: 0.9996 False coordinates of origin: 500,000m Easting; 10,000,

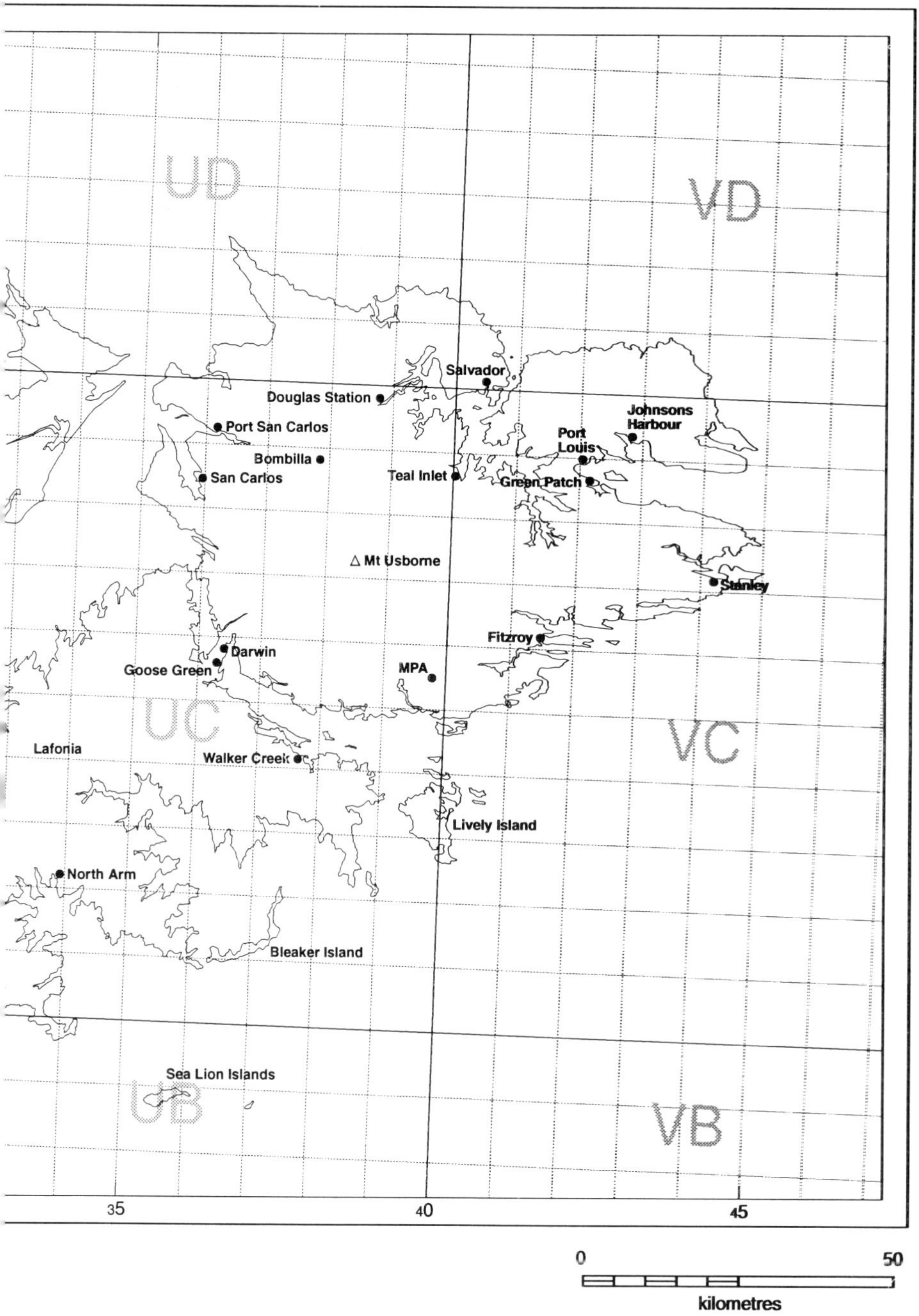
UD
VD
Salvador
Douglas Station
Johnsons Harbour
Port San Carlos
Port Louis
Bombilla
San Carlos
Teal Inlet
Green Patch
△ Mt Usborne
Stanley
Fitzroy
Darwin
Goose Green
MPA
UC
VC
Lafonia
Walker Creek
Lively Island
North Arm
Bleaker Island
Sea Lion Islands
UB
VB
35
40
45
0
50
kilometres

HOSTE INLET [TC4718] After Captain Sir William Hoste Bt. RN (1780-1828). Hoste was Lord Nelson's lieutenant at the Battle of Trafalgar.

HOT STONE COVE CREEK [TD5502] Workers building a fence here at the outbreak of World War II warmed stones in a camp fire and wrapped them in clothing to heat their bedding.

HOTHAM HEIGHTS [TC2046] After Captain Sir Charles Hotham KCB RN (1800-1853) who forced the River Parana in 1845, the brig HMS *Philomel* taking part in the action; he was later Governor of Victoria.

HOUSE COVE, CREEK, ROCKS, etc Descriptive. For example, Black Hill House is visible from the House Rock on Mount Moody.

PORT HOWARD

HOWARD, PORT [UC2375] So called as early as 1766, in all probability after the family of Howard, Earls of Carlisle (see Falkland Sound) by Hon John Byron. It was called Adair's Harbour in the early 18th Century.

HUMMOCK ISLAND [TC6276] POINT [UD1405] Topographical.

HUNSEKER HILL [TC6493] See Huntziker's Leap.

HUNTER'S ARROYO [UC3122] (Spanish) ... Stream.

HUNTZIKER'S LEAP [TC2324] John Frederick Huntziker, a Swiss who had worked for the Patagonian Mission for many years, managed Port Stephens Farm for J.M. Dean. Once, whilst chasing colts, he cleared a 500 feet deep chasm here on horseback. There are several variations of the spelling of his name.

INCA'S POND [UD6816] After an alazan (golden yellow) carthorse of this name who died here.

INLET HILL, POINT, etc Topographical.

INNER HORSE PADDOCK, PASS, etc Descriptive.

IRENE ISLAND [UC9839] After a small local cutter wrecked on Volunteer Point in October 1894.

ISLAND CREEK, HARBOUR, POINT, etc Topographical.

ISLET POINT [VC3676] Topographical. There are islets in the Murrell River here.

ISTHMUS COVE [TC8568] Topographical.

JACK SCOTT MOUNTAIN [UC0572] After a shepherd of this name who worked on the Chartres farm in the early 1900s.

JACKASS RINCON [UD7315] (Spanish) ... Corner. Jackass is the local name for the Magellanic Penguin.

JAMES STREET [Stanley] After Sir James Clark Ross (see Ross Road).

JAMESON'S STREAM [VC3378] After H.B.L. Jameson who was appointed clerk to the Assistant Secretary and Treasurer in 1883.

JASON ISLANDS [TD2030] After HMS *Jason* in which Captain John MacBride sailed to Port Egmont in 1766.

JENESTA POINT [UD1918] See Genesta Point.

JEREMY MOORE AVENUE [Stanley] After General Sir Jeremy Moore, Royal Marines, land force commander during the 1982 war.

JERSEY ROAD [Stanley] The island of Jersey, Channel Islands, donated a substantial sum for the rehabilitation of the Islands after the 1982 war.

JHELUM ROAD [Stanley] After the 428 ton wooden barque that put into Stanley in 1870 in a sinking condition having just rounded Cape Horn. She was condemned and beached at the west end of the harbour. She was used as a hulk, is still in existence and is a familiar landmark.

JIM BIGGS' DITCH [TC8769] After a shepherd of this name. The Biggs family has a long Falkland Island lineage, having arrived in 1842 with Governor Moody.

Pressing wool, Johnson's Harbour

JOCK [UC2791] Probably after Jock Skilling, an early 1900s Port Howard shepherd.

JOHN STREET [Stanley] Perhaps after any of the early explorers who shared this first name: Davis, Byron, MacBride; or John Markham Dean (see Dean Street).

JOHN BISCOE ROAD [Stanley] After the Royal Research Ships (RRS) *John Biscoe* (1st: 1947-55, 2nd: 1956-90), long-serving British Antarctic Survey vessels themselves named after the famous explorer. See Lively Island.

JOHN'S BROOK [UC7292] RINCON [UC7493] After John King (1835-90), a shepherd at Cape Dolphin.

JOHNSON'S HARBOUR [VC2894] After John Johnson of Stanley, a Dane known locally as "Pirate Johnson" supposedly the sole survivor of a treasure burying party in Berkeley Sound. He returned to the Islands in 1841 and died on 30 October 1853, aged 42 or 46.

JOHNSTON'S LEAP [TC9659] In the early 1990s Rod Johnston, a road engineer, took the bend here too fast in his Landrover and came off the road.

JUTLAND PENINSULA [UD0910] After Jutland, Denmark.

KANGAROO VALLEY [TC8267] Driving here tends to be particularly bumpy.

KARINA KIRSTEN ISLAND [UD1609] After a schooner that brought out building materials for the old town hall in Stanley. It was condemned there in 1892 and later beached at Pebble Island.

KELLY ROCKS [VC4774] After the American ship *John R. Kelly*, wrecked on these rocks in 1899.

KELLY'S GARDEN [UC5983] George 'Kelly' Rieves lived at San Carlos and made a garden here in the early 1900s.

KELP BAY, CREEK, LAGOON, etc After the seaweed, locally called Kelp found extensively around the Islands. Anglo-Saxon *culp*: a powder derived from seaweed.

KENT ISLAND [UC1498] MOUNT [VC2374] Named by Captain Robert Fitzroy after Princess Victoria Louisa Mary of Saxe-Coburg-Saalfeld, mother of Queen Victoria and widow of Prince Charles of Leinengen and of Edward Duke of Kent, fourth son of King George III.

KENT ROAD [Stanley] After HMS *Kent* which took part in the Battle of the Falklands 1914.

KEPPEL ISLAND [TD9310] After Augustus, Viscount Keppel (1725-86) who accompanied Anson round the world in 1740-44; a Lord of the Admiralty 1765-66; First Lord of the Admiralty 1768-83; and an Admiral.

KIDNEY COVE [VC4779] ISLAND [VC4880, UC9538] Descriptive of their shape.

King George Bay [TC5777] So called as early as 1797, probably after King George III (1760-1820).

King, Port [UC2546] **Street** [Stanley] After Commander Philip Parker King (1793-1856) captain of the survey sloop HMS *Adventure* whilst surveying Patagonia and Tierra del Fuego, 1826-30.

King's Brook, Creek, Pond, Ridge, etc Probably after shepherds, boundary riders, etc. King's Ridge and Pond, for instance, were named after John King (1835-90), a shepherd at Cape Dolphin.

Kits Creek [TC5109] After an escaped negress slave who was brought to the Islands by sealers. She reputedly lived in a cave and was given meagre food and stores in return for 'services'.

Knob Island, Point, etc Descriptive.

L'antioca Stream, Ridge [UC9457] Corruption of (Spanish) *anteojo*, spyglass... Here the stream used to be curved in an almost perfect circle.

Laberinto, Bahia del [UC6325] Labyrinth Bay. See Adventure Sound.

Ladrillo Island [UC2012] Pronounced locally 'latherijas', (Spanish) Brick ... Captain Smylie (see Smylie Channel) is said to have appropriated the materials landed hereabouts for a house.

Lafonia [UC4040] After an Englishman Samuel Fisher Lafone with business in Montevideo who in 1846 purchased the whole of East Falkland south of the isthmus of Choiseul Sound (i.e. Lafonia). He sold his interests to the Falkland Islands Company in 1851. It was formerly called Rincon del Toro (Spanish - Bull Corner) by gauchos.

Lagoon Bar, etc Descriptive.

Laguna Isla [UC7756] (Spanish) Lake with an Island. Pronounced locally 'Lagan-eyesla', the area was originally known as Mount Misery.

Laguna Seca [UC5061] **Verde** [UC7058] (Spanish) Dry, Green Lake.

Lake Hammond [TC7039] Probably after Robert Hammond (see Hammond Cove).

Lamarche Point [VC3291] After Monsieur Lamarche, Second in Command of the corvette *Uranie* in 1820.

Lamb Marking Pond [UC6863] Lamb marking involves tagging the ears, sorting them into male and female and castrating those rams not wanted for breeding.

Landmark Hill [TC6433] Descriptive.

Landsend Bluff [TC0164] Topographical.

Large Island [UC6321] Descriptive.

Lebon Creek [VC2386] Perhaps after one of de Bougainville's party or a crew member of the *Uranie*, a French corvette wrecked nearby in 1820.

Leicester Creek [TC7239] Perhaps after the breed of sheep.

Leopard Bay [UD4501] **Beach** [TD5509] Probably allusive to the Leopard Seal.

Lerat Point [VC2386] After either one of de Bougainville's men or a crew member of the *Uranie*, a French corvette wrecked nearby in 1820.

Letterbox Hill, Island, Point, etc Tin boxes on legs used to be set up on promontories where boats could land easily to deliver mail, or collect it, from neighbouring islands.

Lewis Mount [TC3920] After a girl, Orissa Lewis, brought up by the owner of Port Stephens, Mr Dean, and his wife.

Limpet Creek [UD9307] Faunistic.

Lion Creek, Point etc Probably allusive to the Sea Lion.

Little Chartres [TC9662] See Chartres Settlement.

Little Creek, Mountain, Pond, etc Descriptive.

Lively Island [UC9835] After the *Lively*, a sealing vessel under the command of John Biscoe who anchored in Berkeley Sound in 1830 on his sealing and exploring expedition which resulted in the discovery of Enderby Land in 1831. Also called Volunteer Island early in the 19th century.

LOAFERS COVE [UC2473] An anchorage suitable for a vessel awaiting a fair wind for Stanley.

LOCH HEAD HOUSE [VD3402] Descriptive.

> **Sheep and Shepherds**
>
> An important influence on the formation of place names, certainly on West Falkland, has been the sheep farming industry. Names such as Dip Paddock, Exmoor Gully, Hoggett Camp and, perhaps, Leicester are all derived thus. However, it is the shepherds themselves who have had the greatest impact. Their hard lives, rich characters and undoubted skills all live on in the names of rocks, hills and other features of the land: 'Su' Binnie's Gate near Fox Bay, Bob Skilling's Island, Davy's Paddock, Duffin's Bridge, Findlay Rocks, Jack Scott Mountain, Jim Bigg's Ditch, King's Brook, Lamosa, McAskills, Miles Creek, Muffler Jack Mountain, Murdo, Perrin's Hole, Stewart's Rock. Their way of life is also reflected in Caravan Paddock, Tent Mountain, Rum Pass and Tuppenny Rock. Indeed, almost every feature on land roamed by sheep had a name or story associated with it. Alas, many have died with those who named them.

LOGGERDUCK POINT [UC0597] Logger is the local name for the common Falklands Flightless Steamer Duck (*Tachyeres brachypterus*).

LONG ISLAND, MOUNTAIN, POINT, etc Descriptive.

LONGDON, MOUNT [VC3275] After Sir James Robert Longden [sic] GCMG (1827-91) who was clerk to the Governor and afterwards Acting Colonial Secretary and a JP. He arrived in the Colony in 1844 and left in January 1862 to become President of the Virgin Islands, West Indies.

LOOKOUT ROCKS [VC4172] Children were rewarded to stand here and look out either for mail boats or ships in need of repair making for Stanley.

LOOP HEAD [TC3157] Properly '*loup*' (French - wolf) head. De Bougainville referred to 'loup-renard' (wolf-fox) in 1771. No doubt this was a reference to the Islands' 'warrah' (*Dusicyon antarcticus*).

LORENZO COTTAGE [UD8307] Perhaps after one Don Lorenzo who lived at the Second Corral according to a 1851 census.

LOTUS ROCK [UD3209] After the Falkland Islands Company mailboat of this name that was wrecked here on 2 October 1872.

LOUIS PASS [VC2875] After Louis Despreaux who took a fall into the stream (Murrell River) here. He arrived in the Falklands in the American whaling brig *Francis* of New London, Connecticut, which was totally wrecked on the north west coast of New Island on 15 February 1852. Despreaux died at Stanley on 11 January 1884 aged 66.

LOUIS, PORT [VC2290] Named Fort St Louis by Louis Antoine de Bougainville in 1764 with reference to King Louis XV of France or perhaps to his own name.

LOW PASS, POINT, etc Probably descriptive, but see Low Bay.

LOW BAY [UC7725], **MOUNT** [VC4279] Properly Lowe. After William Lowe "the son of a respectable land agent in Scotland" (Fitzroy), one of the settlers at Port Louis under Vernet, who escaped from the murderers in 1833. In the same year he sold the sealing schooner *Unicorn* (see Adventure Sound) to Capt Fitzroy who engaged him as a pilot in 1834.

LOWER MALO HOUSE [VC0979] After the Arroyo Malo which is close by. The house exists now as Riverview Farm. Nearby Top Malo House was destroyed in the 1982 war.

LUCAS BAY [TC6515] **RINCON** [TC6317] A whaling ship of this name was wrecked here in the mid-1800s.

LUISA, ESENADA DE [UC9227] (Spanish) Louise Inlet. See Lively Sound.

MACBRIDE HEAD [UD3309] After Captain (later Admiral) John MacBride (1740-1800), Governor at Port Egmont (1766-67) who came out to the Islands in command of HMS *Jason*.

MACKINNON CREEK [UC8449] After Lieutenant (later Captain) Lauchlan Bellingham MacKinnon (born 1815) First Officer under Lieutenant (later Admiral Sir) B.J. Sulivan in the surveying ketch HMS *Arrow* (1838-39). Author of "Some Accounts of the Falkland Islands from a Six Month Residence in 1838-39".

MACKINTOSH'S PASS [UD3501] After John MacKintosh who drowned on the way from Teal Inlet to Estancia in the 1840s.

MAGELLAN COVE [VC3193] After a French whaler the *Magellan* wrecked here on 12/13 January 1832.

MAGELLANIC PENGUINS

MAGGIE ELLIOT ROCK [VC4670] After a ship of that name which was wrecked on the rock 10 September 1875. She was a 790 ton barque of Halifax, Nova Scotia, commanded by John Waters. She was loaded with timber and logwood and was proceeding from Corinto to Hamburg.

MAHOGANY POND [UC5711] A ship's cargo of mahogany was washed up in this area.

MAIN PASSAGE, POINT, etc Descriptive. Allusive either to its importance or to the sea.

MALACARA VALLEY [TC5528] (Spanish) An animal colour - mostly dark with a white crooked blaze down the nose.

MALO CREEK, HILLS, ARROYO, etc See Arroyo Malo.

MANADA PADDOCK [UC0055, TC6450] (Spanish) Herd ... This is where the breeding mares would be kept.

MANYBRANCH HARBOUR [UC3791] Descriptive of the numerous arms of the sea which branch off from the harbour. Also called Hell's Gate and Port Surrey on 18th and 19th century charts.

MAPPA HOUSE [UC4632] Possibly alluding to the Latin word meaning 'flat sheet'. The terrain here is very flat.

MARBLE MOUNTAIN [UD0516] Impressionistic.

MARE HARBOUR [UC9749] Faunistic.

MARGARET HILL [VC3082] Named in 1943 after the barque *Margaret* which arrived in Stanley in 1850 in an unseaworthy condition and is now a submerged hulk at the end of the Dockyard Jetty, Stanley.

MARIA, MOUNT [UC2079] Perhaps after the half-breed Indian woman, Maria, born at Asuncion, Paraguay, who once visited the Falkland Islands in a vessel commanded by Matthew Brisbane. She is frequently mentioned by King and Fitzroy in the "Surveying Voyages of HMS *Adventure* and *Beagle*".

MARKHAM VALLEY [UD1314] After John Markham Dean (see Mount Dean).

MARVILLE BAY [VD1707] (French) Wonder ... So called by de Bougainville in 1764.

MARY HILL [VC4573] A name given in 1935 by Captain C.A.G. Hutchinson RN after Miss Mary Hill, latterly Mrs Duncan Watson, of Stanley. She was a Matron at the King Edward Memorial Hospital. There is a memorial window to her in Stanley Cathedral.

MCASKILLS [UC0065] After Jack McAskill, a Chartres shepherd. His wife, Lottie (Mrs Mac), ran the West Falkland telephone switchboard for many years at their home, Goring House.

MCKAY CLOSE [Stanley] After Sergeant Ian McKay VC of the Parachute Regiment who was killed at Mount Longdon during the 1982 war.

MENGUERA POINT [VC4978] Formerly known as William Point after Port William, and on the plan in the Journal of the Royal Geographical Society 1833 it appears as Cape Pembroke, but this latter is probably a mistake. Mengeary is stated by Dr J.E. Hamilton (Government Naturalist) in 1948 to be a mis-spelling of the South American word "manguera" - horseman (most likely). In his despatch to the Secretary of State No 5/46 of 8 April 1846, Governor Moody describes at length the statements of witnesses in connection with the theft of a boat from Stanley. One statement says that "they took a small boat to go to a place called the Mangera in the neighbourhood of Sparrow Cove to shoot birds and rabbits". Later in the same despatch it describes the "mangera" as a cattle enclosure and says that one John MacKintosh then resided there to look after the cattle. It has been suggested that the name might be a corruption of Menara or Megaera, the vessel in which Lieutenant Drury and the Detachment of Marines reached Stanley in 1858. Possibly after Spanish *manguear*, to round up (cattle) hence an enclosure.

MEREDITH HILL [TC5005], CAPE [TC5104] See Cape Meredith.

MICKEY DOOLAN'S DITCH [TC8242] Michael Doolan was one of two brothers in Fred Cobb's employ for the Falkland Islands Company in the late 1800s.

MIDDLE BAY, CREEK, ISLAND, etc Descriptive.

MILE POND [VC3969] The lake is about a mile long.

MILES CREEK [UC7344] After a Walker Creek shepherd, Thomas Miles, who came to Lafonia from Skerries in the late 1800s.

MILLER'S RIDGE [TC9753] After Sidney Miller, manager at Roy Cove in the 1960s and 70s.

MISERY [TD4202, UC8653] The area is often shrouded in mist or low cloud.

MOCHA MOUNTAIN RIDGE, VALLEY, etc Either from (Spanish) *mocho*, flat-topped - referring to the topography - or after Mocha (Spanish) which are cattle that have been de-horned (polled).

MOFFIT HARBOUR [UC2424] Possibly an American name connected with sealing.

MOLLYMAWK, BIG [TC6019] LITTLE [TC6722] Molly Mawk (or Hawk) is a local name for the black-browed albatross.

MONTEVIDEO, CERRO [UC6992] See Cerro Monte.

MONTY DEAN'S CREEK [VC2186] Montague Dean, owner of Port Stephens in the late 1800s, once lost his boot here.

MOODY BROOK [VC3473], MOUNT [UC0962] STREET [Stanley] After Lieutenant (later Major General) Richard Clement Moody RE (1813-87) who was appointed on the recommendation of Lord Vivian to be Lieutenant Governor and Vice Admiral of the Falkland Islands on 23 August 1841. He arrived at Port Louis in the brig *Hebe* on 15 January 1842, bringing with him the Detachment of the Royal Sappers and Miners. During his tenure as Governor he was promoted Captain RE. He carried out the move of the settlement from Port Louis to Stanley and took up official residence there on 15 July 1844. He retired from the office of Governor in 1847. His brother, the Reverend James Leith Moody BA, who was for many years a naval chaplain, came out to the Islands, the first Colonial Chaplain 1845-53.

MOORE, MOUNT [TC5118] Possibly after Governor Moore.

MORO, THE [UD9301] (Spanish) The Moor. Topographical.

MOTLEY ISLAND [UC9022] POINT [UC8625] Perhaps after one of several officers of the name of Mottley [sic] serving in the Royal Navy during the first half of the 19th century.

MOUND POINT [TC5943] Topographical.

MUDDY CREEK, PASS, POND, etc Descriptive.

MUFFLER JACK MOUNTAIN [UC0875] After Jack McAskill, a shepherd at Chartres in the early 1900s, who always wore a white scarf round his neck.

MULLET CREEK [VC3769, TD9400] Faunistic, the creeks being plentiful fishing places for mullet.

> ### Governor Moody
>
> Arguably the most significant influence on the establishment of a colony and its development was a young Royal Engineer lieutenant, Richard Moody. He was appointed Governor and despatched to the Islands in 1841. The ship in which he came, the *Hebe*, is remembered in the Stanley street of that name. It was Moody, following expert advice and direction, who moved the capital from Port Louis to Jackson Harbour. He called the new settlement Stanley after the 14th Earl of Derby, Secretary of State for the Colonies. Hearnden Water was named after one of two brothers who were in the party. Moody Brook was named at this time as was Murray Heights, after General Murray, Master General of the Ordnance, the area being of great tactical importance. Moody's clerk and storekeeper was a surveyor, Murrell Robinson. He was appointed Surveyor General in 1843 and the main river flowing into Port William bears his name. Many families still living on the Islands can trace their ancestry to the brave and somewhat apprehensive soldiers who formed that first garrison.

MURDO, SOUTH [UC6998], NORTH [UD6903] CAVE/HOUSE ROCKS [UD6804] After Murdo Morrison from Scalpay in the Western Isles of Scotland. He came to the Islands in 1907, aged 19, and was for many years head shepherd at Port San Carlos, where he died in 1961. The Cave/House Rocks were so called because he used to maintain that it would be a good place to stay while shepherding in the area. The North and South Camps were named by Norman Cameron of Port San Carlos in his memory.

MURRAY HEIGHTS [VC4172] In his despatch No 2/1844 relating to the original planning of Port Stanley, Governor Moody states that he had named the high ground south of the town, which he describes as a "military feature of great importance", after General Murray, the Master General of the Ordnance.

MURRELL RIVER [VC3775] After Murrell Robinson who came out to the Islands as a clerk and storekeeper with Governor Moody in 1842. He had been trained as a surveyor and carried out the original survey and layout for the new settlement of Stanley. He was appointed Surveyor General on 1 December 1843. He left the colony in 1846 and was appointed Assistant Surveyor, Cape of Good Hope.

MUSTARD MOUNTAIN [UC9166] Probably a variation of Musters, possibly after Charles Musters, Volunteer First Class in HMS *Beagle* 1831-32; or after Lieutenant George C Musters RN of HMS *Stromboli* who applied for and obtained a lease on 6000 acres in the Colony in 1863.

MUZZIE'S PADDOCK [TD4404] Gladys Napier, mother of Roddy Napier of West Point Island, was known as 'Muzzie'.

NAPIER HILL [TD4305] After the settlement family of this name. Roddy Napier, of West Point Island, is the great-nephew of Arthur Felton.

NARROW POINT, THE NARROWS, etc Topographical.

NAVY POINT [VC4273] Named from the naval reserve on the north side of Stanley Harbour. The land here was reserved for naval purposes by Captain Sir James Clark Ross who commanded the *Erebus* in the British Antarctic Expedition 1839-43. He visited Port William in September 1842 whilst he was at Port Louis, to report on the best site for the settlement deciding in favour of Port William.

NECK [TC5839] Topographical.

NED CASEY'S HILL [TC9385] Travelling from Saunders Island to Hill Cove in the early 1900s, Ned Casey disappeared. His bones were found here 25 years later.

NEEDLES, THE [TD5408] Allusive to the feature of the same name in the Isle of Wight.

NEUQUEN STREAM [UC2377] Probably named by gauchos after the town of this name in central Argentina.

NEW HAVEN, HOUSE, ISLAND, etc Descriptive. New Island was so called by American sealers and whalers from the eastern seaboard of the United States, e.g. New Bedford, Newport, New London.

NAVY POINT

Newing's Creek [VC0695] After a passenger on board a cutter which had been loading the Falkland Islands Company schooner *Thetis*. It was returning to Douglas Station when it sank in Port Salvador in 1893.

Nipple Hill [TC8068] Topographically descriptive.

North Arm [UC3822] Topographical, being the northernmost arm of the sea in the Bay of Harbours.

North Basin, Bluff, Camp, etc Descriptive.

North East Creek, Island, Point, etc Descriptive.

North West Arm, Bay, Islands, etc Descriptive.

Norton Inlet [UC7252] Perhaps a family name in the Sulivan family, e.g. Henry Norton Sulivan.

Old Horse Paddock, House, etc Descriptive.

Orford, Cape See Cape Orford.

Orqueta House [UC4953] (Spanish) Fork ... Properly Horqueta. Descriptive of a fork in the stream.

Outer Arrow Harbour, Black Hill, Island, etc Descriptive.

Owen Roads [UB1797] Possibly after Vice Admiral William Fitzwilliam Owen (1774-1857) who surveyed at the Falkland Islands in 1831 in command of HMS *Eden*.

PACKE'S PORT HOWARD [UC2679] Captain R.C. Packe was a sheep farmer in the Falkland Islands from 1847. The family stayed on the Islands until the Second World War but owned the land until 1980. See Port Howard.

PADDOCKS [VD0102, VC2894] Descriptive.

PALOMA POND [UC5998] SAND BEACH [UD5800] (Spanish) Dove ... Possibly named thus by gauchos mistakenly identifying the all-white Snowy Sheathbill.

PASA LIBRE [various] (Spanish) Free passage. Locally, a pasa libre is a cattle or sheep grid.

PASA MANEAS [various] (Spanish) Hobble (a rope or pair of leather cuffs tying a horse's legs together to prevent it from straying). Horses were hobbled here whilst waiting for the waters to recede thus enabling crossing.

PASSAGE ISLANDS [TC4075] Allusive to the many navigable passages between the islands.

PATTISON, PORT [TD5311] After Captain Pattison who commanded HMS *Carcass*. See Carcass Island.

PEAT BANKS, BOG, ISLAND, etc Descriptive.

PEBBLE ISLAND [UD1815] Descriptive of the translucent pebbles for which the island is noted.

PEMBROKE, CAPE [VC5074] See Cape Pembroke.

PEN POINT [UD2402, TC3734] Faunistic, allusive to swans.

PENARROW POINT [TD6902] Perhaps after Penarrow Point in Cornwall. It was formerly called Campbell's Point after Private John Campbell, one of the original Detachment of Royal Sappers and Miners.

PENGUIN ISLAND [TD7700] Probably faunistic but perhaps after HMS *Penguin*, an armed shallop of 36 tons commanded by Lieutenant Samuel Witterong Clayton and taken out to Port Egmont in the store ship HMS *Endeavour* in 1772-3.

PENGUIN POINT, COVE, ROOKERY, etc Faunistic.

PENN ISLAND [TC1454] After William Penn [1644-1718] of the Society of Friends, or Quakers, who founded Pennsylvania 1680/81. See also Beaver, Coffin, Fox, New, Quaker and Tea Islands and Friend Passage.

PENNY MOUNTAIN [UC0774] Possibly after Richard Penny, mate of the schooner *La Sociedad* which arrived in the Islands in 1832.

PENS POND [UD6714] There are lamb-marking pens nearby.

PERK'S ISLAND [UC1559] Possibly after Robin and Rodney Lee's father who was known as 'Perks' Lee or a Mr Perks who was shepherd at Shag Cove, Port Howard in the early 1900s.

PERRIN'S HOLE [UC3295] A shepherd named Frank Perrin and his family lived at Mount Rosalie House before the Second World War. He was thrown off his horse here in 1937.

PHILIMORE ISLAND [UC9940] Probably after Admiral Sir John Phillimore (1781-1840], Admiral Sir B.J. Sulivan's first commander.

PHILLIP'S POINT [UC1019] Possibly after Captain Phillip Enega, a German from Hamburg who used to anchor off the point. He died at Stanley on 21 December 1911, aged 58.

PHILOMEL PASS [TC6359] MOUNT [TC8960] STREET [Stanley] After HMS *Philomel* under Captain (later Admiral) Sir B. J. Sulivan which took stores to the Islands in 1842 and surveyed there until 1845 when she formed part of the squadron which, under Captain Sir Charles Hotham, forced the River Parana.

PIANO POINT [TD8604] A piano being delivered to Saunders Island in the late 1800s was landed at this point in preference to the bay to the north.

PICASO POND [TC5528] (Spanish) A horse colour - black with a white flash down the nose.

PICKTHORNE, MOUNT [TC5888] After G.W. Pickthorne, clerk in HMS *Philomel*.

PICOS, THE [UC6397] (Spanish) The Peaks.

PIEDRA SOLA [UC7477] (Spanish) Single Stone. So named by South American gauchos.

PIG FLAT, POINT, etc Faunistic.

PIOJO GATE [UC5744] (Spanish) Louse ...

PIONEER ROW [Stanley] A name connected with the party of Greenwich and Chelsea Pensioners who arrived in the Falkland Islands as settlers in 1849. Houses built by them make up most of Drury Street and Pioneer Row.

PITALUGA BAY [UD9604] Pitaluga is a local family name (see Gibraltar ...).

PITT CREEK [TC2050] HEIGHTS [TC2349] ISLAND [TC2052] After William Pitt, Earl of Chatham. See Chatham Harbour.

PLAIN, HOUSE, POND, etc Topographical.

PLATT POINT RINCON [VC0595] Possibly after Mario Platta, an infantry Lieutenant sent by Ruiz Puente in 1769.

PLAYA RIDGE [UD9805] (Spanish) Beach ...

PLEASANT ISLAND, MOUNT, PEAK, POINT, etc So named as early as 1833.

POKE POINT [UC3580] The land here pokes out into the Falkland Sound.

POLINKI POND [VC3999] A polinki (Spanish *palenque*) is a gallows-like structure common on the Islands for butchering beef or sheep.

PONCHOS POND, SHANTY [UC5150] After 'Pancho', who was sent from Hope Place/Saladero in the 1860s with a wheelbarrow and lime to build a small hut or shanty.

POND BAY, PADDOCK, POINT, etc Descriptive.

PONY'S PASS [VC3270] Properly 'Powney's' Pass after Henry Powney who was a gaucho in the Stanley area in 1868.

PORPOISE COVE, CREEK, ISLAND, etc Faunistic.

PRALTOS [UC5364] A corruption of the surname of a gaucho, Dominicus Paralta, who died at Douglas Station in the 1870s.

PRETTY BANKS [VC0186] A surplus of pretty garden plants was planted in these stream banks at one time.

PORT HOWARD

PRINCES STREET [VC2381] A very long and wide stone run named after Princes Street, Edinburgh.

PRONG POINT [VC0327] Descriptive.

PURVIS HOUSE [UC1998] **MOUNT** [UD2300] **PORT** [UD2900] After Rear Admiral John Brett Purvis (born 1787), Commodore and Commander-in-Chief of the squadron on the south-east coast of America in 1842-45.

PYRAMID COVE [UC8937] **POINT** [UC9036] So named as early as 1833.

QUAKER ISLAND [TC2055] Named as early as 1797, after the Quakers. See Penn Island.

QUARK POND [UC2270, UB1096] **ROCKS** [VC0981] Quark is the local name for the Black-crowned Night Heron (*Nycticorax nycticorax cyanocephalus*).

QUEEN CHARLOTTE BAY [TC5050] After the wife of George III.

QUEEN POINT [TC5141] The promontory protrudes into Queen Charlotte Bay.

RACECOURSE [UC2683, UC6590, UC7345] **ROAD** [Stanley] Many racecourses were laid out near to settlements for sport. The racecourse in Stanley is used for the major horse racing event of the Islands.

GOOSE GREEN - HERON - QUARK

Race Point [UD5402, TC2633] After the tide race which exists in the area at certain times.

Racker Pond [TC9650] From Spanish *rajar* - to split. Some split fencing sticks were left here.

Ram Island, Paddock, Point, Gate, etc Faunistic.

Rame Head [TC7599] Possibly corruption of Spanish *ramo* (branch), or after Rame Head, Cornwall.

Rat Castle, Island, etc Faunistic.

Reef Harbour, Island, Point, etc Topographical.

Rees, Mount [TD7502] After John F. Rees, Master of HMS *Philomel* in 1842.

Reservoir Road [Stanley] So named because the reservoir for watering ships was situated at its northern end.

Richard, Mount [TD9708] **Port** [TD5737] After Admiral Sir George Henry Richards KCB FRS (born 1820), a friend of Admiral Sir B.J. Sulivan under whom he served as Lieutenant in HMS *Philomel* in 1842-46.

Rincon ... (Spanish) ... Corner. A piece of land with natural boundaries, often a peninsula, making it an easy place to round up or keep animals.

RIVER CAMP, HARBOUR, ISLAND, etc Topographical.

ROBINSON, MOUNT [TC9779] Probably after William Cleaver Francis Robinson, Governor 1866-70 or possibly after Lieutenant W. Robinson, a commanding officer of the surveying ketch HMS *Arrow*, stationed at the Falkland Islands from about 1839-43.

ROBSON [VC2191] A Tim Robson was shipwrecked in Berkeley Sound in the French whaler *Magellan* in 1832 (see Magellan Cove). He remained on the Islands as a farmer at Port Louis.

ROCK/ROCKY ISLAND, POINT, MOUNTAIN etc Topographical.

RODEO POND [VC3166] A place where cattle were rounded up.

RODNEY BLUFF [TC2424] COVE [TC2525] Probably after Admiral Lord Rodney (1719-92).

ROLON COVE [VC2767] After Thomas Rolon, a native of Montevideo but naturalised in the Islands in 1841, who later lived at Port Louis. Now North Basin.

RORY'S CREEK [UC7045] ISLAND [UC7049] After Rory Findlayson, manager at Walker Creek in the early 1900s or Rory Morrison, shepherd at Darwin.

ROSS ROAD [Stanley] After Sir James Clark Ross (1800-62), who commanded HMS *Erebus* in the British Antarctic Expedition 1839-43 and called at the Islands in 1842.

ROUND HILL, MOUNTAIN, POINT, etc Descriptive.

ROUS CREEK [TC5068] After Admiral the Hon Henry John Rous MP (1795-1877), a Lord of the Admiralty in 1846.

ROY COVE [TC6583] Possibly after ... le Roy, Lieutenant in the *Aigle* (see Eagle Hill) under de Bougainville in 1764; or after Captain Robert Fitzroy (see Fitzroy River, etc); or a Major General William Roy RE (1726-90), a well-known antiquary.

RUANA [UC3446] Either (Spanish) a type of *poncho* (wool cape) or a corruption of *roano* - a roan horse.

RUDD'S PASS [UC7880] After John Rudd, Darwin Camp manager, who was murdered here in 1864.

RUGGLES BAY, ISLAND, RINCON, etc An obscure name connected with sealing and so called as early as 1839.

RUINS POINT [UD3108] There are stone ruins of a building with drainage and sewerage systems, possibly a beef salting works, attributed to Captain Smylie.

Rum Pass [TC8184] Gid McKay, a Chartres shepherd, was carrying two stone jars of rum on his horse in the 1930s having picked them up from Port Howard. His horse fell here and one of the jars broke. The story goes that Gid then drank the stream dry!

Ryan's Cove [TC3332] After a shepherd who worked for Bertrand and Holmested in the late 19th century.

Sabina Point [TC0762] Properly *Sabino*, after a 1225 ton Norwegian barque condemned at Stanley in 1891 after being driven back from Cape Horn. She was towed here in 1910.

Sabruno Ditch [UD9306] A horse colour - mousy. A horse probably died here.

Saddleback [VC3179] Descriptive.

Saddle Up Valley [UD6815] Supposedly the first really sheltered valley after leaving Cape House where one's gear could be tightened if necessary.

Sail Rock [TC4278] From a distance it looks like a sail.

Sal Point [UC9532] Possibly from Spanish *sal* (salt). There may formerly have been salt pans here.

Salt House Point [VC2292] John Whitington established a fish salting plant here in the 1840s.

Saladero [UC5768] (Spanish) Salt Tub. It was here that meat was preserved by salting.

Salinas Beach [UC6257] (Spanish) Saline, i.e. salt water.

Salvador [VD0400] (Spanish) Saviour (Christ).

San Carlos [UC5984], Port [UC6202] After a Spanish sloop the *San Carlos* which visited in May 1768. The Port was originally called San Carlos North but the name was changed at the time of the First World War to avoid confusion with San Carlos (South).

Sand Bay, Creek, Hills, etc Descriptive.

Sapper Hill [VC3871] Commemorative of the Detachment of Royal Sappers and Miners stationed for a long period at the Falkland Islands.

Sarny's Creek [UC6150] The Sarney family lived in Walker Creek.

Saunders Island [TD8307] After Admiral Sir Charles Saunders KB (1713-75). He accompanied Lord Anson round the world 1740-44; First Lord of the Admiralty 1766.

Scott Island [UC6753] Probably after W.J.T. Scott, Second Officer in the ketch HMS *Arrow* (see Sparrow Cove).

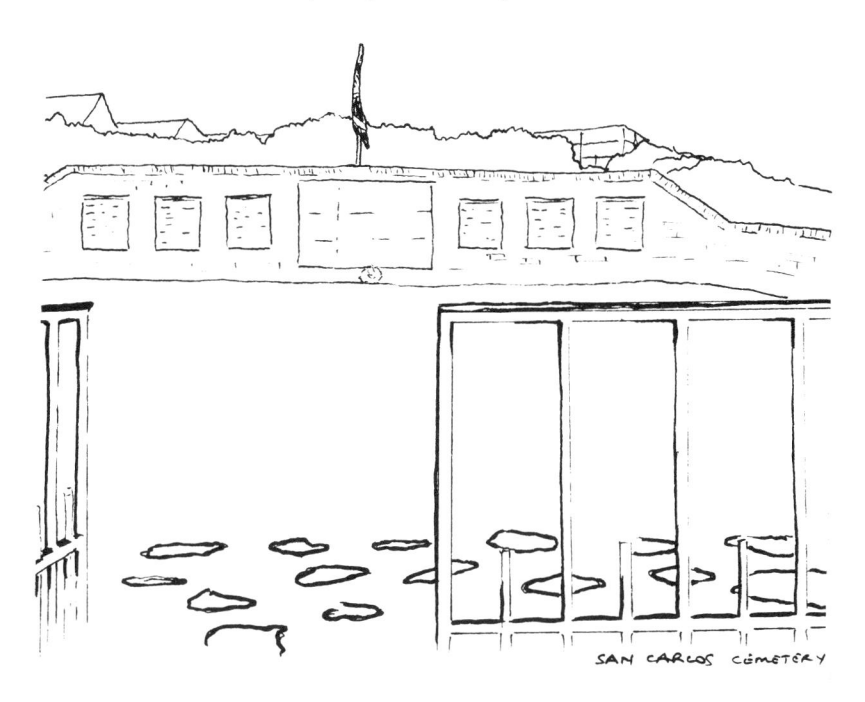

SAN CARLOS CEMETERY

Scoresby Close [Stanley] After the Royal Research Ship *William Scoresby*, based here in the 1930s and 40s. She was named after two men of that name: a whaling captain and his explorer son. The latter published *The Arctic Regions*, a scientific account of the Arctic seas.

Scow Pond [TC9445] A scow is a small barge used to transport wool.

Scrag Paddock [TC7134] Scrags are old sheep for slaughtering.

Sealion Island, Point, Rocks, etc Faunistic.

Seal Cove Island, Point, etc Faunistic.

Second Creek, Island, etc Descriptive.

Semaphore Hill [UC5617, UC7414] Signal boards were erected to attract the attention of and communicate with people on neighbouring islands or the mainland.

SHACKLETON DRIVE [Stanley] After Lord Shackleton (1911-1994), son of Sir Ernest Shackleton the Antarctic explorer. Lord Shackleton's economic reports on the Falkland Islands in 1976 and 1982 had far reaching and beneficial effects on life in the Islands. For many years he was a champion of the Falklands' cause. By his wish, his Garter Banner now hangs in Stanley Cathedral.

SHADROW HILL [UC2284] A shadro(w) is a large saucepan or casserole used for cooking in the embers with an open grate.

SHAG COVE, HARBOUR, ISLANDS, etc Faunistic.

SHALLOP POINT [TC8668] Part of a shallop (French *chaloupe*), a type of boat, was found here.

SHALLOW BAY, BLUFF, HARBOUR, etc Topographical.

SHANTY STREAM [VC2777] So named in 1943 - a shanty standing on the banks of the stream.

SHEDDER POND [VD1104] Geese come here to moult.

SHEILA'S CREEK [VD0701] Correctly Shealers - it was erroneously printed on a 1950s map. A shealer shells corn, etc. or, perhaps, crustaceans.

SHELL ISLAND, POINT, etc Descriptive.

SIGNBOARD HILL [TC4133] A contraption to signal Dyke Island (which has a 'Lookout Hill') was mounted here.

SIMON, MOUNT [UC9478] Pronounced locally Simone. Formerly Mount St Simon after Monsieur N. de St Simon, a Lieutenant of Foot who came to the Islands with de Bougainville in 1766. He was possibly buried here.

SIX HILLS, THE [UC3192] Topographical.

SIXTUS HILL [TC3979] ROCK [VC4695] After a Danish 1707 ton iron barque that was wrecked at the rock on 27 July 1905.

SKIP ROCK [TC8397] From a distance, it looks like a small boat.

SLEIGH VALLEY [UC2493] Loads would be mounted on sleighs which were pulled by horses overland.

SLOOP ROCK [TD4806] From a distance, it looks like a sailing ship.

SMOKO MOUNTAIN [VC1371] Smoko was the local name, of naval origin, for a morning tea/coffee break. The term has since been extended to mean a social gathering.

SMYLIE'S CHANNEL [TC2832] ROCKS [UD6500] VILLAGE [UD6500] After Captain William Horton Smylie, or Smyley, an American sealer and trader who was about the Islands from 1830 onwards.

> ### Captain Smyley
>
> William Horton Smyley (1792-1868) was an entrepreneurial American who developed several careers in the Falkland Islands from 1830 until his death. A cattle and seal hunter, he was also a notorious sea captain. Not averse to piracy, he earned a reputation for policing the seas in the area - for his own good. He was feared by sailors and mistrusted by the British authorities. However, he was not all bad. Whilst carrying out his less legal operations, he also took upon himself to act as coastguard and rescuer of shipwrecked mariners. Among his more altruistic exploits was the search for and discovery of the *Allen Gardiner*, with all crew and passengers lost, in Tierra del Fuego. He brought the vessel back to the Keppel Mission, whence it came. In 1861 he was appointed the United States Consul.

SNAKE HILL [Stanley] A recent name, descriptive of the road's many bends.

SNIPE CAMP [UC6578] Faunistic.

SOUND HOUSE, POINT, RIDGE, etc Topographical.

SPARROW COVE [VC4477] After the ketch HMS *Sparrow*, stationed at the Islands 1839-43. She was commanded first by Lieutenant Lowcay then by Lieutenant Tyssen (see Tyssen Island).

SPEEDWELL ISLAND [UC1610] So named by Lieutenant Edgar 1786-87, most probably after the longboat *Speedwell* built by Commodore Byron after the wreck of the *Wager* (one of Lord Anson's squadron) north of the Straits of Magellan on the coast of Chile in 1741.

SPLIT ISLAND [TC0251, TC4290] Descriptive.

SPRING POINT [TC6353] Topographical - there is a spring in the neighbourhood.

SPUR PADDOCK [TC3823] Topographical.

STAATS ISLAND [TC1243] On HMS *Philomel*'s chart of 1845 it was call Staten Island. This is probably a Dutch or German corruption.

STAG ROAD [VC2689] Probably faunistic. However, as there are no records of stags on the Islands, it is more likely named after a naval surveyor.

STANDING MAN HILL [TC7554, VC 1998] A Standing Man is a rock pillar construction built as a marker for travellers.

STANLEY [VC4172] So named by Governor Moody after Edward Geoffrey Smith Stanley, 14th Earl of Derby (1799-1869), Secretary of State for the Colonies, 1841-44, and Prime Minister in 1852, 1858 and 1866-68. It was Lord Stanley, as he then was, who on the advice of the Admiralty ordered the removal of the chief settlement of the Colony from Port Louis to the Port Jackson area. Stanley Harbour was previously called 'Jackson Harbour', probably after Andrew Jackson, President of the USA 1829-37, Port William then being a harbour frequented by the American whalers who visited the Islands in the first half of the 19th century. Earlier the French under de Bougainville had named it Beau Port.

Murder

A small but highly significant event in the history of the Islands occurred at Port Louis in 1833 and is remembered in the street name Brisbane Road, Stanley. Matthew Brisbane was in charge of the Port Louis Settlement at the time when he was cruelly murdered by a band of convicts and gauchos. In 1864, at Rudd's Pass north of Mount Usborne, John Rudd, manager at Darwin Camp, was murdered by Gill, a half-breed Indian gaucho, according to contemporary accounts.

STARVATION PEAK [UC9275] The vegetation here is sparse, therefore holding little of sustenance for animals.

STATES COVE [TC3149] Formerly States Bay. So named by Captain Benjamin Hussey of the sealing vessel *United States*, probably in allusion to the numerous US sealers and whalers which frequented the West Falklands in the 18th and 19th centuries. Captain Portlock calls the principal harbour in States Bay Hussey's Harbour in honour of the discoverer.

STEEPLE JASON [TD0537] See Jason Islands.

STEPHENS, PORT [TC3821] So called as early as 1766 after Sir Philip Stephens Bart FRS (1725-1809), Second Secretary of the Admiralty 1759-62, First Secretary 1763-95 and Lord of the Admiralty 1795-1806.

STEVELLY HILL [TC6194] Probably after Stevely, a midshipman in HMS *Philomel*.

STEWART'S ROCK [UC1772] Gordon Stewart, a shepherd from Port Howard, used to shelter with his horse at this large rock.

STINKER ISLAND [UC0921, TC1150] Stinker is the local name for the Southern Giant Petrel (*Macronectes giganteus*).

ST MARY'S WALK [Stanley] After St Mary's Roman Catholic Chapel.

STONE, STONERUN, STONEY PASS, etc Topographical.

STRANGER'S HOLE [UC3998] At some time in the 1940s a horse called Stranger was injured here when he became entangled in his tack.

STRAWBERRY HILL [UC7834] After the wild strawberry (*Rubus geoides*) found on the Islands whose fruit resembles that of a raspberry.

STRIKE OFF POINT [VC3284] The point for which vessels made when sailing up Berkeley Sound from which to 'strike off' in order to clear Long Island.

STUD FLOCK, PADDOCK, POINT, etc Descriptive.

STURGESS POINT [UD2008] Possibly after a mariner, who was part of the Keppel Island Mission, shipwrecked here.

SUFFOLK HILL [UC1874] Perhaps after the breed of sheep.

SUGAR LOAF [TC6088, TC5097] After the mountain of this name at the entrance to Rio de Janiero harbour. The latter was a significant feature, it marking the first/last port of call for travellers leaving/returning to the Falkland Islands.

SULIVAN HARBOUR [UC5131] MOUNT [TC5085] After Admiral Sir Bartholomew James Sulivan (1810-90); a Lieutenant in HMS *Beagle*; commanded the survey ships HMS *Arrow* in 1838 and *Philomel* in 1862 at the Falkland Islands. His eldest son, Commander James Young Falkland Sulivan, was the first British subject to be born at Port Louis.

SURF BAY [VC4672] Topographical.

SUSSEX [UC6375] Probably allusive to the English county. So named as early as 1840, it was the port for the area before San Carlos was established.

SWAN INLET, ISLAND, PASSAGE, etc Faunistic.

SYMONDS HARBOUR [TC6952] Perhaps after Rear Admiral Sir William Symonds KCB FRS (1782-1836), Surveyor of the Navy 1832-47.

TAMAR PASS [UD3109], CAPE [UD2715] So named by Hon John Byron after the sloop HMS *Tamar*, Commanded by Captain Patrick Mouat, which accompanied the *Dolphin* to the Falkland Islands in 1765.

TAM'S ROCK [UD6501] After Tam Sheddon who worked at Port San Carlos in the 1920s-40s.

TEA ISLAND [TC1342] Possibly named by Quakers and alluding to the Boston Tea Party.

TEAL CREEK, INLET, RIVER, etc Faunistic.

CARAVAN - TEAL INLET

TELEPHONE GATE [VC0562] HILL [TC6689] The old system of magneto telephones required wires carried on poles. These features would be identified as being near to where such wires were situated.

TEN SHILLING BAY [TC4512] Sealers often used to bet between themselves on spontaneous races to shore after catching trips. The prize would be in this case ten shillings.

TENT MOUNTAIN [UC1278] PASS [UC4034] Referring to shelters made by shepherds and boundary riders with low stone walls covered with canvas (probably from old sails) held down by stones.

TERN HILL, POINT, etc Faunistic.

TERRA MOTAS POINT [UC5875] (Spanish) Flat or Bog ...

TERRIBLE, CAPE [TD4007] Descriptive.

THATCHER DRIVE [Stanley] After Margaret (now Baroness) Thatcher, Prime Minister during the 1982 war.

THETIS BAY [UC3921] After the Falkland Islands Company schooner that was lost with all hands and without trace off the north coast of the Islands in July 1901.

TIDE ISLAND, POINT, etc Descriptive.

TINWHISTLE GATE [UC4538] In the late 19th century some gate posts were made from two metal halves through which the wind whistled. Many are still in existence.

TOP COOKHOUSE [TC9564] The old settlement of Chartres used to be located around here prior to the move to its current position at the end of the nineteenth century.

TORCIDA POINT [UC7644] (Spanish) Twisted ...

TOWN POINT [TC7672] Named during the voyage of HMS *Philomel*. Numerous sandy patches on the hill give the appearance of houses.

TRANQUILIDAD [UC5361] (Spanish) Peacefulness. Pronounced locally "Trinkly-dar". Once a shepherd's house, it is now a remote ruin.

TRISTE ISLAND [UC8322] Possibly after the Spanish word meaning 'sad'.

TRYPOT [UD7409] Iron cauldrons called trypots or tripots were used by sealers to boil down the carcasses of seals and penguins to extract the valuable oil.

TUMBLEDOWN MOUNTAIN [VC3372] Should be "The Tumbledown"; it commemorates a herd of horses which was driven over the steep rock face of the hill by gauchos.

TUPPENNY ROCK [UC0763] Somebody once lost two pence here. It was a common place for shepherds to meet to receive instructions from the Chartres farm manager.

TURKEY ISLAND [TC7080] ROCKS [UC1093] Faunistic, referring to the Turkey Vulture (*Cathartes aura falklandica*).

TURNER'S STREAM [VC3280] After Private Samuel Turner of the Detachment of Royal Sappers and Miners.

TUSSAC ISLAND, POINT, ROCKS, etc Tussac or tussock (*Parodiochloa flabellata*) is a common, tall grass around the coast.

TWELVE O'CLOCK MOUNTAIN [VC3980] HILL [VC2297] So called because they are due north of Government House, Stanley and Port Louis Settlement respectively.

TWO BOB VALLEY [UC0479] Possibly attributed to Tommy "Two Bob" Skilling. Two bob refers to two shillings.

TWO SISTERS, THE [VC3073] Descriptive of two peaks that from a distance look the same height.

TYSSEN ISLANDS [UC1848] After Lieutenant John Tyssen RN, Officer in Charge at Port Louis, 1839-42. He commanded the ketch HMS *Sparrow* (see Sparrow Cove).

URANIE BAY [VC2685] After the French frigate *Uranie* which in 1820, on a scientific expedition led by L de Freycinet, received damage in Le Maire Straits. De Freycinet decided to take refuge at the Falkland Islands, but the ship struck a rock off Volunteer Rocks, drifted up Berkeley Sound and finally beached in this bay.

USBORNE, MOUNT [UC7371] After Alexander Burns Usborne, Master's Assistant in HMS *Beagle* 1831-36.

UPLAND GEESE

VAMPIRE BEACH [UC7228] After a schooner lost here during December 1870.

VERDE, CAMPO [UC6387], **INNER** [UC6387] **MOUNTAINS** [UC6086] (Spanish) Green ...

VERNET, MOUNT [VC2480] After Louis Vernet, of French birth, a merchant of Hamburg. A South American by naturalisation, he took an expedition to East Falkland in the *Alerte* in 1826. He was appointed Governor of Malvinas and Tierra del Fuego by the government of Buenos Aires in 1828 but his settlement at Port Louis was destroyed in 1831 by the US ship *Lexington* after he had arrested the American sealing schooners *Breakwater*, *Harriet* (see Mount Harriet) and *Superior*.

VERONICA POINT [TC5952] After the native 'Boxwood' plant (*Hebe elliptica*) commonly grown as hedging.

VICTORIA HARBOUR [UC7445] So named as early as 1839 after Queen Victoria.

VILLIERS STREET [Stanley] After Edward Villiers, one of the Colonial Land and Emigration Commissioners in 1843.

VIPER, THE [VC5074] A rock named after HMS *Viper* which visited the Islands in 1884.

VIXEN'S DITCH [TC9398] This is reputed to be the place where the last Falkland Islands fox (warrah) was killed.

VOLUNTEER POINT [VC4992] After the ship *Volunteer* in which Captain Edmund Fanning (see Fannings Head) called at Port Louis in 1815.

WAGGON POINT [TC9899] The wheels came off a wagon at this point at one time when the two trace horses pulling it bolted.

WALKER CREEK [UC7741] Obscure but so named by 1843.

WALL MOUNTAIN [VC2770] So named in 1943, descriptive of the rock face which forms a long, high ridge.

WARRAH BRIDGE, HOUSE, RIVER, etc Allusive to the Falkland Island quadruped (*Dusicyon antarcticus*) called by early settlers the 'warrah'.

WATERFALL CREEK, MOUNTAIN, STREAM, etc Topographical.

WATT COVE [VC3974] Properly Watts Cove. So named after Private William Watts of the Detachment of Sappers and Miners. Watts was afterwards a Corporal and a Police Constable. Also a Lieutenant J.S. Watts of HMS *Narcissus* surveyed Stanley Harbour in 1868.

WEASEL'S BAY [TC9038] After a boat wrecked in the vicinity.

WEDDELL ISLAND [TC2543] SETTLEMENT [TC3144] After Captain James Weddell RN (1787-1834) Antarctic explorer, etc; the discoverer of the Weddell Sea. In the 18th and early 19th centuries it was called Swan Island.

WEDGE ISLAND [UC0633] Descriptive of its shape.

WEIR CREEK [VC4075] The site of a fish weir.

WEST POINT ISLAND [TD4303] Presumably so called because it lies off one of the westerly points of the Islands.

WETHER GROUND [various] Where the castrated rams (wethers) are kept.

WHALE BAY, ISLAND, PASSAGE, etc Faunistic.

WHALER BAY [TC5885] Descriptive. It was a favourite resort of whalers and sealers in the early 19th century.

WHARTON HARBOUR [UC2439] Perhaps after Commander Francis Wharton RN who died in 1848, or his sons John Antony Lawrence and Richard Hill Wharton, both RN.

WHIG ISLANDS [UC7348] Possibly allusive to the 'Whig' political party.

WHINBUSH BAY [UC8835] Otherwise gorse (*Ulex europaeus*), an introduced plant grown extensively for hedging.

WHITE ROCK [UD4703] Descriptive. A prominent white rock, the subject of comment by Captain Woodes Rogers in 1708 and by many others.

WHITINGTONS RINCON [VC1786] Possibly after the brothers G.T. and J.B. Whitington who produced papers and a prospectus on the benefits of farming and fishing in the Falkland Islands *circa* 1828-50.

WICKHAM HEIGHTS [UC8967], MOUNT [UC9467] After Lieutenant (later Captain) John Clements Wickham RN, Lieutenant in HMS *Adventure* (see Adventure Sound) 1827-30, and First Lieutenant in HMS *Beagle* 1831-36, surveying at the Falkland Islands 1833-34. He later became Governor of Queensland.

WILLIAM, PORT [VC4575] MOUNT [VC3471] Named by Captain Robert Fitzroy after William, Duke of Clarence (third son of King George III) who reigned as William IV 1830-37. Port William was also known as Port Firm in 1838 from its being a sheltered harbour with good holding ground for anchors. Still earlier it was called Baye Choiseul (see Choiseul Sound) by the French under de Bougainville.

WILLIAMS' ROADWAY [TC5530] After H. Williams who owned Port Edgar in the late 19th century.

WINDY RIDGE [TC9772] Descriptive.

WINE, WINEGLASS POINT, etc Descriptive of the feature's shape.

WIRELESS RIDGE [VC3774] A powerful naval and government wireless station was located below this ridge at the head of Stanley Harbour between 1917-20.

WOOLLY GUT, THE [TD4505] Named originally by Ernest Holmested, a West Falkland farmer. It is a contraction of 'williwaw' - a sudden wind from high ground.

Wrecks

Sadly, many a name is derived from the wrecks of the great number of merchant ships that either served the Islands or foundered there after abortive attempts to round Cape Horn. The wind and shores were unforgiving and, in those early pioneering days, navigation was hazardous. De Bougainville's corvette '*Uranie*' was destroyed in the bay at the head of Berkeley Sound that now bears her name. Large trading vessels such as the 790-ton '*Maggie Elliott*' (1875), the German '*Concordia*' (1891), the '*John R Kelly*' (1899), in 1905, the '*Sixtus*' (1707 tons), and the Norwegian barque '*Bertha*', with a cargo of cedar, all ended their days on the Islands' coasts. The whaling ships '*Magellan*' in 1832 and '*Lucas*' in the mid-19th century both met a similar fate. Some smaller, inter-island boats were also casualties. The Falkland Islands Company '*Lotus*' hit the rock that now bears her name in 1872; '*Annie Brooks*' perished near Fox Bay in April 1874; '*Genesta*' was damaged at Port Egmont in 1888 and subsequently lost at sea; '*Don Carlos*', a local cutter, foundered in 1889; the schooner '*Foam*', broke up on Carcass Island in 1890. The names of all these wrecks live on.

WRECK ISLANDS [TD7227] After a Spanish snow, a boat similar to a brig (a Dutch word), wrecked off the islands.

WRINKLY HILL [TC9244] Descriptive.

YATES' VALLEY [VC2291] Thomas Yates, a bricklayer, his wife and three children arrived at Port Louis on 15 January 1842. He was a Private in Governor Moody's party of Royal Sappers and Miners.

YORKE BAY [VC4573] Probably named after Charles Philip Yorke, fourth Earl of Hardwicke (1799-1872) who served for many years in the Royal Navy, becoming an admiral in 1863. The name may have been given by Fitzroy.

YOUNG, MOUNT [TC5423] Probably after the family to which Sir B.J. Sulivan's wife, a daughter of Vice Admiral James Young, belonged.

ZAINO RINCON [VC1086] (Spanish) Chestnut Horse Corner.

ROCKHOPPER PENGUINS